Computational Intelligence Methods and Applications

Series Editors

Sanghamitra Bandyopadhyay, Machine Intelligence Unit, Indian Statistical Institute, Kolkata, West Bengal, India

Ujjwal Maulik, Department of Computer Science & Engineering, Jadavpur University, Kolkata, West Bengal, India

Patrick Siarry, LiSSi, E.A. 3956, Université Paris-Est Créteil, Vitry-sur-Seine, France

The monographs and textbooks in this series explain methods developed in computational intelligence (including evolutionary computing, neural networks, and fuzzy systems), soft computing, statistics, and artificial intelligence, and their applications in domains such as heuristics and optimization; bioinformatics, computational biology, and biomedical engineering; image and signal processing, VLSI, and embedded system design; network design; process engineering; social networking; and data mining.

More information about this series at https://link.springer.com/bookseries/15197

Bahram Farhadinia

Hesitant Fuzzy Set

Theory and Extension

Springer

Bahram Farhadinia
Department of Mathematics
Quchan University of Technology
Quchan, Iran

ISSN 2510-1765 ISSN 2510-1773 (electronic)
Computational Intelligence Methods and Applications
ISBN 978-981-16-7303-0 ISBN 978-981-16-7301-6 (eBook)
https://doi.org/10.1007/978-981-16-7301-6

This Springer imprint is published by the registered company Springer Nature Singapore Pte Ltd.
The registered company address is: 152 Beach Road, #21-01/04 Gateway East, Singapore 189721, Singapore

To my wife, Saeedeh, *who never stopped believing in me, and to my sons,* Babak *and* Bardia, *who shine brighter than the sun…*

Preface

Hesitant fuzzy set has moved beyond its infancy and is now entering a maturing phase with increased numbers and types of extensions. Up to now, the existing extensions of hesitant fuzzy set have been encountering an increasing interest and attracting more and more attentions. It is not an exaggeration if it is said that the recent decade has seen the blossoming of a larger set of techniques and theoretical outcomes for hesitant fuzzy sets and their extensions as well as applications.

Inevitably, researches on hesitant fuzzy set and its extensions have inspired a huge amount of publications, including a considerable number of high-impact journal contributions, numerous edited volumes, and monographs. This is while, in spite of the existence of many good monographs, creating a comprehensive book that covers and introduces all kinds of exiting extensions of hesitant fuzzy sets in a unified framework has remained a little neglected or ignored. Thus, it is becoming more and more necessary to gather and provide an organized and rather unified view of these extensions and their key notions, with reference to the relevant literature.

According to the feedback from several experts and based on my own experience gained through the past ten years, the main motivation of writing this book is to provide the students, scholars, and interested readers with a monograph that covers all kinds of exiting hesitant fuzzy extensions in a unified scheme. Further, I hope that this all-encompassing reference book enables scientists and engineers to understand the central aspects of hesitant fuzzy extensions.

This book gives preliminary, but fundamental, information that may be useful for a better understanding of hesitant fuzzy extensions in a unified framework. The current book is organized into twelve chapters that deal with twelve different but related issues, which are listed as follows:

Chapter 1 provides the readers with further background on hesitant fuzzy set as the core concept of this book, and fundamental information bases for the study of direct generalized forms of hesitant fuzzy set. In this regard, the basic operational laws, fundamental definitions, basic operations, different kinds of negations, S-norms, and T-norms together with two kinds of ordering techniques for hesitant fuzzy sets are presented in the first section of this chapter. The other section of this chapter deals with the direct extensions of hesitant fuzzy sets just by representing their basic concepts in brief together with their corresponding set or algebraic operations. Most popular among these direct generalizations are the hesitant triangular fuzzy set, interval-valued hesitant fuzzy set, extended hesitant fuzzy set, higher order of hesitant fuzzy set, dual hesitant fuzzy set, dual hesitant triangular fuzzy set, and interval-valued dual hesitant fuzzy set.

Chapter 2, in the first section, initially introduces the concept of hesitant fuzzy linguistic term set that reflects the inconsistency and uncertainty of experts. Then, the extended form of hesitant fuzzy linguistic term set, namely extended hesitant fuzzy linguistic term set, is re-stated and a number of its operations are reviewed. In the sequel, the interval-valued hesitant fuzzy linguistic term set is taken into consideration in the third section. In the fifth and fourth sections, the concepts of proportional hesitant fuzzy linguistic term set and hesitant fuzzy uncertain linguistic set are, respectively, addressed. The concept of dual hesitant fuzzy linguistic term set, which allows us to take much more information into account, is reviewed in the sixth section. Two other generalized forms of hesitant fuzzy linguistic term set, which are known as dual hesitant fuzzy linguistic triangular set and interval-valued dual hesitant fuzzy linguistic set, are the issues that lie at the end arguments in this chapter.

Chapter 3 is devoted to introducing briefly the concept of neutrosophic set in the first section. Subsequently, the concept of single-valued neutrosophic hesitant fuzzy set, which is refereed here to as neutrosophic hesitant fuzzy set, is addressed in the second section. In the sequel, it is going to argue about the concept of interval neutrosophic hesitant fuzzy set. Although there may be reported some other kinds of neutrosophic hesitant fuzzy set by different researchers, the aforementioned two types of neutrosophic-based hesitant fuzzy sets are enough to fully understand the other kinds.

Chapter 4 focuses on Pythagorean hesitant fuzzy set in the first section. Next, in order to put forward the issues and concepts described in this chapter, an overall view of interval-valued form of Pythagorean hesitant fuzzy set is given in the second section. The last concept that is needed for the complete characterization of Pythagorean hesitant fuzzy set is that examined here as the dual Pythagorean hesitant fuzzy set.

Chapter 5 deals with q-rung orthopair hesitant fuzzy set in its first section. The concept of interval-valued q-rung orthopair hesitant fuzzy set is the next issue that will be put forward in the second section of this chapter. The last concept, which is needed for the complete characterization of q-rung orthopair hesitant fuzzy set, is the dual q-rung orthopair hesitant fuzzy set.

Chapter 6, in the first section, presents the core of the next concepts, known as probabilistic hesitant fuzzy set. Then, the dual form of probabilistic hesitant fuzzy sets is given in the second section. The third section of this chapter is dedicated to introduce the notion of occurring probability of possible values into hesitant fuzzy linguistic term set and define probabilistic linguistic hesitant fuzzy set. Keeping the implication of probabilistic linguistic term set in mind, it is concerned in the fourth section with the definition of probabilistic linguistic dual hesitant fuzzy set by combining probabilistic dual hesitant fuzzy set with dual hesitant fuzzy linguistic term set. In the later sections, the other generalizations of probabilistic hesitant fuzzy set, namely interval probabilistic hesitant fuzzy linguistic variable, probabilistic neutrosophic hesitant fuzzy set, Pythagorean probabilistic hesitant fuzzy set, and q-rung orthopair probabilistic hesitant fuzzy set, will be introduced, and some of their corresponding operations will be given.

Chapter 7 focuses firstly on the notion of type 2 hesitant fuzzy set. Then, the interval-valued type 2 hesitant fuzzy set, which is also known as interval type 2 hesitant fuzzy set, will be given in the last section of this chapter.

Chapter 8 provides the interested reader with the concept of hesitant bipolar fuzzy set in the first section. The second section deals with the concept of hesitant bipolar-valued fuzzy set that comes from bipolar concept, and it is stated more or less different from the concept of hesitant bipolar fuzzy set. In the subsequent section of this chapter, the concept of hesitant bipolar-valued neutrosophic set will be addressed.

Chapter 9 introduces the concept of cubic hesitant fuzzy set and then presents the form in which cubic hesitant fuzzy set is characterized in the form of triangle, and hence it is known as triangular cubic hesitant fuzzy set. In the next section of this chapter, the definition of linguistic intuitionistic cubic hesitant variable will be stated.

Chapter 10 includes a study of complex hesitant fuzzy set together with complex dual hesitant fuzzy set. This chapter contains the basic key points of complex dual type 2 hesitant fuzzy set.

Chapter 11 presents the concept of picture hesitant fuzzy set and, then, introduces the concept of interval-valued picture hesitant fuzzy set. The last section of this chapter is devoted to presenting the concept of dual picture hesitant fuzzy set.

Chapter 12 focuses on the concept of spherical hesitant fuzzy set as a generalization of both picture hesitant fuzzy set and Pythagorean hesitant fuzzy set for dealing with fuzziness and uncertainty information. In the sequel, a generalized form of spherical hesitant fuzzy set, known as T-spherical hesitant fuzzy set, will be given.

In this book, an attempt to review the researches previously conducted on the latest baselines of hesitant fuzzy sets and their extensions is made in order for helping the interested researchers and practitioners get informed about the fundamental concepts and also the corresponding references on the relevant research in this field.

Quchan, Iran Bahram Farhadinia
May 2021

Contents

Acronyms

Chapter 1

A	Hesitant fuzzy set (HFS)
$\underline{A}$	Hesitant triangular fuzzy set (HTFS)
$\tilde{A}$	Interval-valued hesitant fuzzy set (IVHFS)
$\grave{A}$	Extended hesitant fuzzy set (EHFS)
$\breve{A}$	Higher order hesitant fuzzy set (HOHFS)
$\mathbf{A}$	Dual hesitant fuzzy set (DHFS)
$\bar{A}$	Dual hesitant triangular fuzzy set (DHTFS)
$\tilde{\mathbf{A}}$	Interval-valued dual hesitant fuzzy set (IVDHFS)

Chapter 2

A	Hesitant fuzzy linguistic term set (HFLTS)
$\tilde{A}$	Interval-valued hesitant fuzzy linguistic term set (IVHFLTS)
$\underline{A}$	Proportional hesitant fuzzy linguistic term set (PHFLTS)
$\grave{A}$	Hesitant fuzzy uncertain linguistic set (HFULS)
$\mathbf{A}$	Dual hesitant fuzzy linguistic set (DHFLS)
$\bar{A}$	Dual hesitant fuzzy triangular linguistic set (DHFTLS)
$\tilde{\mathbf{A}}$	Interval-valued dual hesitant fuzzy linguistic set (IVDHFLS)

Chapter 3

A	Neutrosophic hesitant fuzzy set (NHFS)
$\tilde{A}$	Interval neutrosophic hesitant fuzzy set (INHFS)

Chapter 4

A	Pythagorean hesitant fuzzy set (PHFS)
$\widetilde{A}$	Interval-valued Pythagorean hesitant fuzzy set (IVPHFS)
$\mathbf{A}$	Dual Pythagorean hesitant fuzzy set (DPHFS)

Chapter 5

A	q-Rung orthopair hesitant fuzzy set (q-ROHFS)
$\widetilde{A}$	Interval-valued q-rung orthopair hesitant fuzzy set (IVq-ROHFS)

Chapter 6

A	Probabilistic hesitant fuzzy set (PHFS)
$\widetilde{A}$	Probabilistic dual hesitant fuzzy set (PDHFS)
$\mathbf{A}$	Probabilistic linguistic hesitant fuzzy set (PLHFS)
$\widetilde{\mathbf{A}}$	Probabilistic linguistic dual hesitant fuzzy set (PLDHFS)
$\underline{\underline{A}}$	Interval probability hesitant fuzzy linguistic variable (IPHFLV)
$\overline{A}$	Probabilistic neutrosophic hesitant fuzzy set (PNHFS)
$\acute{A}$	Pythagorean probabilistic hesitant fuzzy set (PPHFS)
$\grave{A}$	q-Rung orthopair probabilistic hesitant fuzzy set (q-ROPHFS)

Chapter 7

A	Type 2 hesitant fuzzy set (T2HFS)
$\widetilde{A}$	Interval type 2 hesitant fuzzy set (IT2HFS)

Chapter 8

A	Hesitant bipolar fuzzy set (HBFS)
$\widetilde{A}$	Hesitant bipolar-valued fuzzy set (HBVFS)
$\mathbf{A}$	Hesitant bipolar-valued neutrosophic set (HBVNS)

Chapter 9

A	Cubic hesitant fuzzy set (CHFS)
$\widetilde{A}$	Triangular cubic hesitant fuzzy set (TCHFS)
$\mathbf{A}$	Linguistic intuitionistic cubic hesitant variable (LICHV)

Chapter 10

A	Complex hesitant fuzzy set (CHFS)
$\widetilde{A}$	Complex dual hesitant fuzzy set (CDHFS)
$\mathbf{A}$	Complex dual type 2 hesitant fuzzy set (CDT2HFS)

Chapter 11

A	Picture hesitant fuzzy set (PHFS)
$\widetilde{A}$	Interval-valued picture hesitant fuzzy set (IVPHFS)

Chapter 12

A	Spherical hesitant fuzzy set (SHFS)
$\widetilde{A}$	T-spherical hesitant fuzzy set (T-SHFS)

Chapter 1
Hesitant Fuzzy Set

Abstract In this chapter, we first introduce the core concept of this book, known as the concept of hesitant fuzzy set. By providing the reader with the next part of the book, we are going to look at the basic topics of hesitant fuzzy sets, including fundamental definitions, basic operations, different kinds of negations, S-norms and T-norms together with two kinds of ordering techniques for hesitant fuzzy sets. In the sequel, we deal with a class of direct generalizations of hesitant fuzzy set, just by representing their concepts briefly together with their corresponding set or algebraic operations. Most popular among these direct generalizations are the hesitant triangular fuzzy set, interval-valued hesitant fuzzy set, extended hesitant fuzzy set, higher order of hesitant fuzzy set, dual hesitant fuzzy set, dual hesitant triangular fuzzy set, and interval-valued dual hesitant fuzzy set.

1.1 Hesitant Fuzzy Set

In the case where different sources of vagueness appear simultaneously, the concept of fuzzy set [58] is not able to properly model the uncertainty, imprecise, and vague information. In order to overcome such a limitation, different type of extensions of fuzzy set have been introduced in the literature. Among such extensions, we may mention the intuitionistic fuzzy set [3] allowing to be considered simultaneously the membership degree and the non-membership degree of each element, type-2 fuzzy set [11] incorporating the concept of uncertainty in the definition of membership function where a fuzzy set over [0, 1] is used to model it, interval-valued fuzzy set [48] assigning to each element a closed subinterval of [0, 1] as the membership degree of that element, fuzzy multiset [38] being based on multiset with possibly repeated elements.

(continued)

B. Farhadinia, *Hesitant Fuzzy Set*, Computational Intelligence Methods and Applications, https://doi.org/10.1007/978-981-16-7301-6_1

However, a more widely and frequently used concept of fuzzy extensions is hesitant fuzzy set which is introduced by Torra [47] for modelling a situation of decision making in which an expert might consider different degrees of membership of an element in a set.

Among the pioneer researchers, who provided a broad conceptualization and application of hesitant fuzzy set and was followed by different researchers on the area later, were Torra [47], Xu [52], Herrera [44], Wei [50], Liao [35], and Farhadinia [13].

A HFS is semantically described by a function that returns a set of membership degrees for any element in the domain, and it is given as below.

Definition 1.1 ([47]) Let X be a reference set. A hesitant fuzzy set (HFS) on X is described by the function

$$A : X \to \wp([0, 1]), \tag{1.1}$$

which returns a non-empty subset of values in [0, 1].

Here, the notation of $\wp([0, 1])$ indicates the non-empty subset of values in [0, 1].

Remark 1.1 Throughout this book, the set of all HFSs on the reference set X is denoted by $\mathbb{HFS}(X)$.

However, Xia and Xu [54] put forwarded the seminal definition of HFS with the following easier mathematical representation:

Definition 1.2 ([54]) Let X be a fixed reference set. A HFS on X is defined in terms of a function from X to a subset of [0, 1] which is characterized by

$$A = \{\langle x, h_A(x)\rangle \mid x \in X\}, \tag{1.2}$$

in which $h_A(x)$ is called a hesitant fuzzy element (HFE) and it denotes a set of some values in [0, 1]. Moreover, the HFE $h_A(x)$ stands for the possible membership degrees of the element $x \in X$ to the set A.

Based on the above notation, we easily find that the HFS $A = \bigcup_{x \in X}\{h_A(x)\}$ can be defined in terms of a set of all HFEs of A in which the HFEs is considered with finite number of elements [5].

Example 1.1 If $X = \{x_1, x_2, x_3\}$ is the reference set, $h_A(x_1) = \{0.2, 0.5, 0.6\}$, $h_A(x_2) = \{0.1, 0.4\}$, and $h_A(x_3) = \{0.1, 0.2, 0.5, 0.7\}$ are the HFEs of x_i $(i = 1, 2, 3)$ to a set A, respectively. Then A can be considered as a HFS, i.e.,

$$A = \{\langle x_1, \{0.2, 0.5, 0.6\}\rangle, \langle x_2, \{0.1, 0.4\}\rangle, \langle x_3, \{0.1, 0.2, 0.5, 0.7\}\rangle\}.$$

Taking the reference set X into account, the following HFEs are known as the special HFEs:

- Empty set: $h(x) = \{0\} \triangleq O^*$;
- Full set: $h(x) = \{1\} \triangleq I^*$;
- Complete ignorance: $h(x) = [0, 1] \triangleq U^*$;
- Nonsense set: $h(x) = \{\,\} \triangleq \emptyset^*$.

In order to deal with HFEs, Torra [47] introduced initially several basic operations as the following:

Definition 1.3 ([47]) Let h_A be a HFE, its complement is defined as:

$$h_A^c = \bigcup_{\gamma_A \in h_A} \{1 - \gamma_A\}. \tag{1.3}$$

Definition 1.4 ([47]) Let h_{A_1} and h_{A_2} be two HFEs, their union and intersection are, respectively, defined as follows:

$$h_{A_1} \cup h_{A_2} = \bigcup_{\gamma_{A_1} \in h_{A_1}, \gamma_{A_2} \in h_{A_2}} \max\{\gamma_{A_1}, \gamma_{A_2}\}, \tag{1.4}$$

$$h_{A_1} \cap h_{A_2} = \bigcup_{\gamma_{A_1} \in h_{A_1}, \gamma_{A_2} \in h_{A_2}} \min\{\gamma_{A_1}, \gamma_{A_2}\}. \tag{1.5}$$

From a mathematical point of view, a HFE can be seen as the other well-known extensions of fuzzy sets where for any $x \in X$:

- If $h_A(x) = \{\gamma_A\}$ with $\gamma_A \in [0, 1]$, then the HFE $h_A(x)$ is a fuzzy set;
- If $h_A(x) = \{\gamma_A^1, \gamma_A^2, \ldots, \gamma_A^n\} \subseteq [0, 1]$ with $\gamma_A^i \neq \gamma_A^j$ for $i \neq j$, then the HFE $h_A(x)$ is a multiset in which every membership is different from each other [4];
- If $h_A(x) = [\underline{h}_A(x), \overline{h}_A(x)] \subseteq [0, 1]$, then the HFE $h_A(x)$ is an interval-valued fuzzy set [48];
- If $h_A(x) = [\underline{h}_A(x), \overline{h}_A(x)] \subseteq [0, 1]$, then by considering the mathematical equivalence between interval-valued fuzzy sets and intuitionistic fuzzy sets [54], the HFE $h_A(x)$ is an intuitionistic fuzzy set.

For three HFEs h_A, h_{A_1}, and h_{A_2}, Xia and Xu [53] re-defined the following HFE arithmetic operations:

$$h_{A_1} \oplus h_{A_2} = \bigcup_{\gamma_{A_1} \in h_{A_1}, \gamma_{A_2} \in h_{A_2}} \{\gamma_{A_1} + \gamma_{A_2} - \gamma_{A_1} \times \gamma_{A_2}\}; \tag{1.6}$$

$$h_{A_1} \otimes h_{A_2} = \bigcup_{\gamma_{A_1} \in h_{A_1}, \gamma_{A_2} \in h_{A_2}} \{\gamma_{A_1} \times \gamma_{A_2}\}; \tag{1.7}$$

$$\lambda h_A = \bigcup_{\gamma_A \in h_A} \{1 - (1 - \gamma_A)^\lambda\},\ \lambda > 0; \quad (1.8)$$

$$h_A^\lambda = \bigcup_{\gamma_A \in h_A} \{\gamma_A^\lambda\},\ \lambda > 0. \quad (1.9)$$

Note that the definition of set operations on HFEs is not the only possible expressions which are described by Definitions 1.3 and 1.4. More complete classes of expressions dedicated to the HFE set operations are those given by the classes of negations, S-norms (or T-conorms), and T-norms. However, the above-mentioned set operations on HFEs are called the standard fuzzy operations.

In order to produce a family of meaningful fuzzy operations on fuzzy sets or their extensions, we consider the following axiomatic requirements [33]:

Negation/Complement Axioms

- N1. $N(0) = 1$ and $N(1) = 0$ (Boundary conditions);
- N2. For all $a, b \in [0, 1]$, if $a \leq b$, then $N(a) \geq N(b)$ (Monotonicity);
- N3. For each $a \in [0, 1]$, $N(N(a)) = a$ (Involutivity),

where $N : [0, 1] \times [0, 1] \to [0, 1]$.

T-norm/Intersection Axioms

- T1. $T(a, 1) = a$ (Boundary condition);
- T2. For all $a, b, c \in [0, 1]$, if $b \leq c$, then $T(a, b) \leq T(a, c)$ (Monotonicity);
- T3. For all $a, b \in [0, 1]$, $T(a, b) = T(b, a)$ (Commutativity);
- T4. For all $a, b, c \in [0, 1]$, $T(a, T(b, c)) = T(T(a, b), c)$ (Associativity),

where $T : [0, 1] \times [0, 1] \to [0, 1]$.

S-norm/Union Axioms

- S1. $S(a, 0) = a$ (Boundary condition);
- S2. For all $a, b, c \in [0, 1]$, if $b \leq c$, then $S(a, b) \leq S(a, c)$ (Monotonicity);
- S3. For all $a, b \in [0, 1]$, $S(a, b) = S(b, a)$ (Commutativity);
- S4. For all $a, b, c \in [0, 1]$, $S(a, S(b, c)) = S(S(a, b), c)$ (Associativity),

where $S : [0, 1] \times [0, 1] \to [0, 1]$.

We call a T-norm function $T : [0, 1] \times [0, 1] \to [0, 1]$ as an Archimedean T-norm if it is continuous and satisfies $T(a, a) < a$, and moreover, we call a S-norm function $S : [0, 1] \times [0, 1] \to [0, 1]$ as an Archimedean S-norm if it is continuous and satisfies $S(a, a) > a$.

The theorem of T-norm/S-norm characterization [33] enables us to generate a class of Archimedean T-norms/S-norms by using the non-unique *additive generator* g as follows:

$$T(a, b) = g^{-1}(g(a) + g(b)), \quad (1.10)$$

and similarly, an Archimedean S-norm by using the non-unique additive generator k as the following:

$$S(a, b) = k^{-1}(k(a) + k(b)), \tag{1.11}$$

in which $g : [0, 1] \to [0, \infty)$ is a strictly decreasing function such that $g(1) = 0$, and $k(a) = g(1 - a)$.

Furthermore, an Archimedean T-norm may be generated by the non-unique *multiplicative generator* $\hat{g}$ in the form of

$$T(a, b) = \hat{g}^{-1}(\hat{g}(a)\hat{g}(b)), \tag{1.12}$$

where $\hat{g}(a) = exp(-g(a))$ for all $a \in [0, 1]$.

In what follows, we present some T-norms (respectively, S-norms) which are known as Algebraic, Einstein, Hamacher, and Frank T-norms (respectively, S-norms) based on choosing different additive generators g and k [6]:

- If $g(a) = -log(a)$, and therefore $k(a) = -log(1 - a)$, $g^{-1}(a) = e^{-a}$, $k^{-1}(a) = 1 - e^{-a}$, then one gets Algebraic T-norm and Algebraic S-norm as:

$$T_1(a, b) = ab, \tag{1.13}$$

$$S_1(a, b) = a + b - ab. \tag{1.14}$$

- If $g(a) = log(\frac{2-a}{a})$, and therefore $k(a) = log(\frac{2-(1-a)}{1-a})$, $g^{-1}(a) = \frac{2}{e^a+1}$, $k^{-1}(a) = 1 - \frac{2}{e^a+1}$, then one gets Einstein T-norm and Einstein S-norm as:

$$T_2(a, b) = \frac{ab}{1 + (1 - a)(1 - b)}, \tag{1.15}$$

$$S_2(a, b) = \frac{a + b}{1 + ab}. \tag{1.16}$$

- If $g(a) = log(\frac{\epsilon+(1-\epsilon)a}{a})$, $\epsilon > 0$, and therefore $k(a) = log(\frac{\epsilon+(1-\epsilon)(1-a)}{1-a})$, $g^{-1}(a) = \frac{\epsilon}{e^a+\epsilon-1}$, $k^{-1}(a) = 1 - \frac{\epsilon}{e^a+\epsilon-1}$, then one gets Hamacher T-norm and Hamacher S-norm as:

$$T_3^{\epsilon}(a, b) = \frac{ab}{\epsilon + (1 - \epsilon)(a + b - ab)}, \tag{1.17}$$

$$S_3^{\epsilon}(a, b) = \frac{a + b - ab - (1 - \epsilon)ab}{1 - (1 - \epsilon)ab}, \quad \epsilon > 0. \tag{1.18}$$

- If $g(a) = log(\frac{\epsilon-1}{\epsilon^a-1})$, $\epsilon > 1$, and therefore $k(a) = log(\frac{\epsilon-1}{\epsilon^{1-a}-1})$, $g^{-1}(a) = \frac{log(\frac{\epsilon-1+e^a}{e^a})}{log(\epsilon)}$, $k^{-1}(a) = 1 - g^{-1}(a) = 1 - \frac{log(\frac{\epsilon-1+e^a}{e^a})}{log(\epsilon)}$, then one gets Frank T-norm and Frank S-norm as:

$$T_4^{\epsilon}(a,b) = log_{\epsilon}\left(1 + \frac{(\epsilon^a - 1)(\epsilon^b - 1)}{\epsilon - 1}\right), \tag{1.19}$$

$$S_4^{\epsilon}(a,b) = 1 - log_{\epsilon}\left(1 + \frac{(\epsilon^{1-a} - 1)(\epsilon^{1-b} - 1)}{\epsilon - 1}\right), \quad \epsilon > 1. \tag{1.20}$$

Definition 1.5 ([17]) Given two HFEs represented by $h_{A_1} = \bigcup_{\gamma_{A_1} \in h_{A_1}} \{\gamma_{A_1}\}$ and $h_{A_2} = \bigcup_{\gamma_{A_2} \in h_{A_2}} \{\gamma_{A_2}\}$. Several product and sum operations on the HFEs which are also HFEs can be described as follows:

- Algebraic product and Algebraic sum:

$$\begin{aligned} h_{A_1} \otimes h_{A_2} &= \bigcup_{\gamma_{A_1} \in h_{A_1}, \gamma_{A_2} \in h_{A_2}} \{T_1(\gamma_{A_1}, \gamma_{A_2})\} \\ &= \bigcup_{\gamma_{A_1} \in h_{A_1}, \gamma_{A_2} \in h_{A_2}} \{\gamma_{A_1}\gamma_{A_2}\}, \end{aligned} \tag{1.21}$$

$$\begin{aligned} h_{A_1} \oplus h_{A_2} &= \bigcup_{\gamma_{A_1} \in h_{A_1}, \gamma_{A_2} \in h_{A_2}} \{S_1(\gamma_{A_1}, \gamma_{A_2})\} \\ &= \bigcup_{\gamma_{A_1} \in h_{A_1}, \gamma_{A_2} \in h_{A_2}} \{\gamma_{A_1} + \gamma_{A_2} - \gamma_{A_1}\gamma_{A_2}\}; \end{aligned} \tag{1.22}$$

- Einstein product and Einstein sum:

$$\begin{aligned} h_{A_1} \otimes h_{A_2} &= \bigcup_{\gamma_{A_1} \in h_{A_1}, \gamma_{A_2} \in h_{A_2}} \{T_2(\gamma_{A_1}, \gamma_{A_2})\} \\ &= \bigcup_{\gamma_{A_1} \in h_{A_1}, \gamma_{A_2} \in h_{A_2}} \left\{\frac{\gamma_{A_1}\gamma_{A_2}}{1 + (1 - \gamma_{A_1})(1 - \gamma_{A_2})}\right\}, \end{aligned} \tag{1.23}$$

$$\begin{aligned} h_{A_1} \oplus h_{A_2} &= \bigcup_{\gamma_{A_1} \in h_{A_1}, \gamma_{A_2} \in h_{A_2}} \{S_2(\gamma_{A_1}, \gamma_{A_2})\} \\ &= \bigcup_{\gamma_{A_1} \in h_{A_1}, \gamma_{A_2} \in h_{A_2}} \left\{\frac{\gamma_{A_1} + \gamma_{A_2}}{1 + \gamma_{A_1}\gamma_{A_2}}\right\}; \end{aligned} \tag{1.24}$$

- Hamacher product and Hamacher sum ($\epsilon > 0$):

$$h_{A_1} \otimes h_{A_2} = \bigcup_{\gamma_{A_1} \in h_{A_1}, \gamma_{A_2} \in h_{A_2}} \{T_3^{\epsilon}(\gamma_{A_1}, \gamma_{A_2})\}$$
$$= \bigcup_{\gamma_{A_1} \in h_{A_1}, \gamma_{A_2} \in h_{A_2}} \left\{ \frac{\gamma_{A_1}\gamma_{A_2}}{\epsilon + (1-\epsilon)(\gamma_{A_1} + \gamma_{A_2} - \gamma_{A_1}\gamma_{A_2})} \right\}, \tag{1.25}$$

$$h_{A_1} \oplus h_{A_2} = \bigcup_{\gamma_{A_1} \in h_{A_1}, \gamma_{A_2} \in h_{A_2}} \{S_3^{\epsilon}(\gamma_{A_1}, \gamma_{A_2})\}$$
$$= \bigcup_{\gamma_{A_1} \in h_{A_1}, \gamma_{A_2} \in h_{A_2}} \left\{ \frac{\gamma_{A_1} + \gamma_{A_2} - \gamma_{A_1}\gamma_{A_2} - (1-\epsilon)\gamma_{A_1}\gamma_{A_2}}{1 - (1-\epsilon)\gamma_{A_1}\gamma_{A_2}} \right\}; \tag{1.26}$$

- Frank product and Frank sum ($\epsilon > 1$):

$$h_{A_1} \otimes h_{A_2} = \bigcup_{\gamma_{A_1} \in h_{A_1}, \gamma_{A_2} \in h_{A_2}} \{T_4^{\epsilon}(\gamma_{A_1}, \gamma_{A_2})\}$$
$$= \bigcup_{\gamma_{A_1} \in h_{A_1}, \gamma_{A_2} \in h_{A_2}} \left\{ log_{\epsilon}(1 + \frac{(\epsilon^{\gamma_{A_1}} - 1)(\epsilon^{\gamma_{A_2}} - 1)}{\epsilon - 1}) \right\}, \tag{1.27}$$

$$h_{A_1} \oplus h_{A_2} = \bigcup_{\gamma_{A_1} \in h_{A_1}, \gamma_{A_2} \in h_{A_2}} \{S_4^{\epsilon}(\gamma_{A_1}, \gamma_{A_2})\}$$
$$= \bigcup_{\gamma_{A_1} \in h_{A_1}, \gamma_{A_2} \in h_{A_2}} \left\{ 1 - log_{\epsilon}(1 + \frac{(\epsilon^{1-\gamma_{A_1}} - 1)(\epsilon^{1-\gamma_{A_2}} - 1)}{\epsilon - 1}) \right\}. \tag{1.28}$$

Now, we turn to the following definitions which reveal the axiomatic skeleton properties of fuzzy division and fuzzy subtraction (see [17]).

Division Axioms

- T^*1. $T^*(a, 1) = a$ (Boundary condition);
- T^*2. For all $a, b, c \in [0, 1]$, if $b \leq c$, then $T^*(b, a) \leq T^*(c, a)$ (Monotonicity);
- T^*3. For all $a \in [0, 1]$, $T^*(a, a) > a$ (Super-idempotency),

where $T^* : [0, 1] \times [0, 1] \to [0, 1]$.

Subtraction Axioms

- S^*1. $S^*(a, 0) = a$ (Boundary condition);
- S^*2. For all $a, b, c \in [0, 1]$, if $b \leq c$, then $S^*(b, a) \leq S^*(c, a)$ (Monotonicity);
- S^*3. For all $a \in [0, 1]$, $S^*(a, a) < a$ (Sub-idempotency),

where $S^* : [0, 1] \times [0, 1] \to [0, 1]$.

Farhadinia [17] proved that for the Archimedean T-norm $T : [0, 1] \times [0, 1] \to [0, 1]$ and the Archimedean S-norm $S : [0, 1] \times [0, 1] \to [0, 1]$, which are expressed by their additive generators in the forms of

$$T(a,b) = g^{-1}(g(a) + g(b)), \tag{1.29}$$

$$S(a,b) = k^{-1}(k(a) + k(b)), \tag{1.30}$$

we can derive the following results:

$$T(a,\overline{a}) = 1, \quad \text{if and only if} \quad g(\overline{a}) = -g(a), \tag{1.31}$$

$$S(a,\overline{a}) = 0, \quad \text{if and only if} \quad k(\overline{a}) = -k(a). \tag{1.32}$$

Here, $g : [0,1] \rightarrow [0,\infty)$ is a strictly decreasing function such that $g(1) = 0$, and $k(a) = g(1-a)$.

On the basis of the above results, we can express the operational functions T^*, S^* with respect to the Archimedean T-norm $T : [0,1] \times [0,1] \rightarrow [0,1]$ and the Archimedean S-norm $S : [0,1] \times [0,1] \rightarrow [0,1]$ as follows:

$$T^*(a,b) := T(a,\overline{b}) = g^{-1}(g(a) - g(b)), \tag{1.33}$$

$$S^*(a,b) := S(a,\overline{b}) = k^{-1}(k(a) - k(b)). \tag{1.34}$$

Hereafter, we call the functions g and k as the *subtractive generators* of operational functions T^*-norm and S^*-norm, respectively.

Also, a T^*-norm can be expressed by its *divisive generator* $\hat{g}$ as:

$$T^*(a,b) = \hat{g}^{-1}\left(\frac{\hat{g}(a)}{\hat{g}(b)}\right), \tag{1.35}$$

where $\hat{g}(a) = exp(g(a))$ for all $a \in [0,1]$.

Notice that the subtractive generator of T^*-norm (respectively, S^*-norm) is indeed the additive generator of T-norm (respectively, S-norm), and thus the subtractive generator of T^*-norm inherits all the properties of the additive generator of T-norm.

Now, we are interested in presenting a fundamental definition which provides us with a method for producing several division and subtraction operators on HFEs.

Definition 1.6 ([17]) Let $C[0,1]$ be a set of some values in $[0,1]$, and $T^*, S^* : [0,1] \times [0,1] \rightarrow [0,1]$ be operational functions given by (1.33) and (1.34). Then, the division and subtraction operators on $C[0,1]$ are, respectively, defined as:

$$\oslash(a,b) = \min\{1, T^*(a,b)\} = \min\{1, g^{-1}(g(a) - g(b))\}, \tag{1.36}$$

$$\ominus(a,b) = \max\{0, S^*(a,b)\} = \max\{0, k^{-1}(k(a) - k(b))\}, \tag{1.37}$$

where $g : [0,1] \rightarrow [0,\infty)$ is a strictly decreasing function such that $g(1) = 0$, and $k(a) = g(1-a)$.

In the following, we introduce some T^*-norms and S^*-norms that are obtained in accordance with different subtractive generators g and k:

- If $g(a) = -log(a)$, and therefore $k(a) = -log(1-a)$, $g^{-1}(a) = e^{-a}$, $k^{-1}(a) = 1 - e^{-a}$, then one gets Algebraic T^*-norm and Algebraic S^*-norm as:

$$T_1^*(a,b) = \frac{a}{b}, \tag{1.38}$$

$$S_1^*(a,b) = \frac{a-b}{1-b}. \tag{1.39}$$

- If $g(a) = log(\frac{2-a}{a})$, and therefore $k(a) = log(\frac{2-(1-a)}{1-a})$, $g^{-1}(a) = \frac{2}{e^a+1}$, $k^{-1}(a) = 1 - \frac{2}{e^a+1}$, then one gets Einstein T^*-norm and Einstein S^*-norm as:

$$T_2^*(a,b) = \frac{a(2-b)}{1-(1-a)(1-b)}, \tag{1.40}$$

$$S_2^*(a,b) = \frac{a-b}{1-ab}. \tag{1.41}$$

- If $g(a) = log(\frac{\epsilon+(1-\epsilon)a}{a})$, $\epsilon > 0$, and therefore $k(a) = log(\frac{\epsilon+(1-\epsilon)(1-a)}{1-a})$, $g^{-1}(a) = \frac{\epsilon}{e^a+\epsilon-1}$, $k^{-1}(a) = 1 - \frac{\epsilon}{e^a+\epsilon-1}$, then one gets Hamacher T^*-norm and Hamacher S^*-norm as:

$$T_3^{*\epsilon}(a,b) = \frac{(\epsilon+(1-\epsilon)b)a}{\epsilon b-(1-\epsilon)(a-b-ab)}, \tag{1.42}$$

$$S_3^{*\epsilon}(a,b) = \frac{a-b}{(1-\epsilon)((1-a)(1-b)-(1-a))+(1-b)}, \quad \epsilon > 0. \tag{1.43}$$

- If $g(a) = log(\frac{\epsilon-1}{\epsilon^a-1})$, $\epsilon > 1$, and therefore $k(a) = log(\frac{\epsilon-1}{\epsilon^{1-a}-1})$, $g^{-1}(a) = \frac{log(\frac{\epsilon-1+e^a}{e^a})}{log(\epsilon)}$, $k^{-1}(a) = 1 - g^{-1}(a) = 1 - \frac{log(\frac{\epsilon-1+e^a}{e^a})}{log(\epsilon)}$, then one gets Frank T^*-norm and Frank S^*-norm as:

$$T_4^{*\epsilon}(a,b) = log_\epsilon\left(1+(\epsilon-1)\frac{\epsilon^a-1}{\epsilon^b-1}\right), \tag{1.44}$$

$$S_4^{\epsilon}(a,b) = 1 - log_\epsilon\left(1+(\epsilon-1)\frac{\epsilon^{1-a}-1}{\epsilon^{1-b}-1}\right), \quad \epsilon > 1. \tag{1.45}$$

In the sequel, we are in a position to construct several division and subtraction operations for HFEs based on the above-defined T^*-norm and S^*-norm:

Definition 1.7 ([17]) Given two HFEs represented by $h_{A_1} = \bigcup_{\gamma_{A_1} \in h_{A_1}} \{\gamma_{A_1}\}$ and $h_{A_2} = \bigcup_{\gamma_{A_2} \in h_{A_2}} \{\gamma_{A_2}\}$. Several division and subtraction operations on the HFEs which are also HFEs can be described as follows:

- Algebraic division and Algebraic subtraction:

$$h_{A_1} \oslash h_{A_2} = \bigcup_{\gamma_{A_1} \in h_{A_1}, \gamma_{A_2} \in h_{A_2}} \oslash(\gamma_{A_1}, \gamma_{A_2})$$
$$= \bigcup_{\gamma_{A_1} \in h_{A_1}, \gamma_{A_2} \in h_{A_2}} \min\{1, T_1^*(\gamma_{A_1}, \gamma_{A_2})\}\}$$
$$= \bigcup_{\gamma_{A_1} \in h_{A_1}, \gamma_{A_2} \in h_{A_2}} \min\left\{1, \frac{\gamma_{A_1}}{\gamma_{A_2}}\right\}, \tag{1.46}$$
$$h_{A_1} \ominus h_{A_2} = \bigcup_{\gamma_{A_1} \in h_{A_1}, \gamma_{A_2} \in h_{A_2}} \ominus(\gamma_{A_1}, \gamma_{A_2})$$
$$= \bigcup_{\gamma_{A_1} \in h_{A_1}, \gamma_{A_2} \in h_{A_2}} \max\{0, S_1^*(\gamma_{A_1}, \gamma_{A_2})\}\}$$
$$= \bigcup_{\gamma_{A_1} \in h_{A_1}, \gamma_{A_2} \in h_{A_2}} \max\left\{0, \frac{\gamma_{A_1} - \gamma_{A_2}}{1 - \gamma_{A_2}}\right\}; \tag{1.47}$$

- Einstein division and Einstein subtraction:

$$h_{A_1} \oslash h_{A_2} = \bigcup_{\gamma_{A_1} \in h_{A_1}, \gamma_{A_2} \in h_{A_2}} \oslash(\gamma_{A_1}, \gamma_{A_2})$$
$$= \bigcup_{\gamma_{A_1} \in h_{A_1}, \gamma_{A_2} \in h_{A_2}} \min\{1, T_2^*(\gamma_{A_1}, \gamma_{A_2})\}\}$$
$$= \bigcup_{\gamma_{A_1} \in h_{A_1}, \gamma_{A_2} \in h_{A_2}} \min\left\{1, \frac{\gamma_{A_1}(2 - \gamma_{A_2})}{1 - (1 - \gamma_{A_1})(1 - \gamma_{A_2})}\right\}, \tag{1.48}$$
$$h_{A_1} \ominus h_{A_2} = \bigcup_{\gamma_{A_1} \in h_{A_1}, \gamma_{A_2} \in h_{A_2}} \ominus(\gamma_{A_1}, \gamma_{A_2})$$
$$= \bigcup_{\gamma_{A_1} \in h_{A_1}, \gamma_{A_2} \in h_{A_2}} \max\{0, S_2^*(\gamma_{A_1}, \gamma_{A_2})\}\}$$
$$= \bigcup_{\gamma_{A_1} \in h_{A_1}, \gamma_{A_2} \in h_{A_2}} \max\left\{0, \frac{\gamma_{A_1} - \gamma_{A_2}}{1 - \gamma_{A_1}\gamma_{A_2}}\right\}; \tag{1.49}$$

- Hamacher division and Hamacher subtraction for $\epsilon > 0$:

$$h_{A_1} \oslash h_{A_2} = \bigcup_{\gamma_{A_1}\in h_{A_1},\gamma_{A_2}\in h_{A_2}} \oslash(\gamma_{A_1},\gamma_{A_2})$$
$$= \bigcup_{\gamma_{A_1}\in h_{A_1},\gamma_{A_2}\in h_{A_2}} \min\{1, T_3^{*\epsilon}(\gamma_{A_1},\gamma_{A_2})\}$$
$$= \bigcup_{\gamma_{A_1}\in h_{A_1},\gamma_{A_2}\in h_{A_2}} \min\left\{1, \frac{(\epsilon+(1-\epsilon)\gamma_{A_2})\gamma_{A_1}}{\epsilon\gamma_{A_2}-(1-\epsilon)(\gamma_{A_1}-\gamma_{A_2}-\gamma_{A_1}\gamma_{A_2})}\right\}, \tag{1.50}$$

$$h_{A_1} \ominus h_{A_2} = \bigcup_{\gamma_{A_1}\in h_{A_1},\gamma_{A_2}\in h_{A_2}} \ominus(\gamma_{A_1},\gamma_{A_2})$$
$$= \bigcup_{\gamma_{A_1}\in h_{A_1},\gamma_{A_2}\in h_{A_2}} \max\{0, S_3^{*\epsilon}(\gamma_{A_1},\gamma_{A_2})\}$$
$$= \bigcup_{\gamma_{A_1}\in h_{A_1},\gamma_{A_2}\in h_{A_2}} \max\left\{0, \frac{\gamma_{A_1}-\gamma_{A_2}}{(1-\epsilon)((1-\gamma_{A_1})(1-\gamma_{A_2})-(1-\gamma_{A_1}))+(1-\gamma_{A_2})}\right\}; \tag{1.51}$$

- Frank division and Frank subtraction for $\epsilon > 1$:

$$h_{A_1} \oslash h_{A_2} = \bigcup_{\gamma_{A_1}\in h_{A_1},\gamma_{A_2}\in h_{A_2}} \oslash(\gamma_{A_1},\gamma_{A_2})$$
$$= \bigcup_{\gamma_{A_1}\in h_{A_1},\gamma_{A_2}\in h_{A_2}} \min\{1, T_4^{*\epsilon}(\gamma_{A_1},\gamma_{A_2})\}$$
$$= \bigcup_{\gamma_{A_1}\in h_{A_1},\gamma_{A_2}\in h_{A_2}} \min\left\{1, log_\epsilon\left(1+(\epsilon-1)\frac{\epsilon^{\gamma_{A_1}}-1}{\epsilon^{\gamma_{A_2}}-1}\right)\right\}, \tag{1.52}$$

$$h_{A_1} \ominus h_{A_2} = \bigcup_{\gamma_{A_1}\in h_{A_1},\gamma_{A_2}\in h_{A_2}} \ominus(\gamma_{A_1},\gamma_{A_2})$$
$$= \bigcup_{\gamma_{A_1}\in h_{A_1},\gamma_{A_2}\in h_{A_2}} \max\{0, S_4^{*\epsilon}(\gamma_{A_1},\gamma_{A_2})\}$$
$$= \bigcup_{\gamma_{A_1}\in h_{A_1},\gamma_{A_2}\in h_{A_2}} \max\left\{0, 1-log_\epsilon\left(1+(\epsilon-1)\frac{\epsilon^{1-\gamma_{A_1}}-1}{\epsilon^{1-\gamma_{A_2}}-1}\right)\right\}. \tag{1.53}$$

Notice that HFEs should be considered as a set of some values in the open-interval (0, 1) to avoid zero denominator when calculating the above operational functions.

It is worthwhile to mention that the above-proposed Algebraic division and Algebraic subtraction are in accordance with the only ones introduced by Liao and Xu in [36].

However, Farhadinia [17] generalized the definitions of division and subtraction operations in Definition 1.6 in the forms of

$$\oslash(a, b) = \min\{1, T^*(a, b)\} = \min\{1, g^{-1}(g(a) - g(b))\}, \tag{1.54}$$

$$\ominus(a, b) = \max\{0, S^*(a, b)\} = \max\{0, k^{-1}(k(a) - k(b))\}. \tag{1.55}$$

In the latter formulas $g : [0, 1] \rightarrow [0, \infty)$ stands for a strictly decreasing function such that $g(1) = 0$, and $k(a) = g(1 - a)$.

Moreover, if we consider the negation operator $N(a) = 1 - a$, then the pair of division and subtraction operations satisfy the following De Morgans laws:

$$\oslash(a, b) = N(\ominus(N(a), N(b))), \tag{1.56}$$

$$\ominus(a, b) = N(\oslash(N(a), N(b))). \tag{1.57}$$

Comparison of two HFEs is one of the most frequently used tool in decision making technique. However, the number of values in different HFEs may be different. To have a correct comparison, the two HFEs $h_{A_1}(x)$ and $h_{A_2}(x)$ should have the same length l_x. Suppose that $l(h_{A_1}(x))$ stands for the number of values in $h_{A_1}(x)$. In such a situation, we make the assumptions that: (see [50, 52, 54, 64]) (A1) All the elements in each $h_{A_1}(x)$ are arranged in increasing order, and then $\gamma_{A_1}^{\sigma(j)}(x)$ is referred to as the j-th largest value in $h_{A_1}(x)$. (A2) If, for some $x \in X$, $l(h_{A_1}(x)) \neq l(h_{A_2}(x))$, then $l_x = \max\{l(h_{A_1}(x)), l(h_{A_2}(x))\}$. If there are fewer elements in $h_{A_1}(x)$ than in $h_{A_2}(x)$, an extension of $h_{A_1}(x)$ should be considered by repeating its maximum element or its minimum element or any element until it has the same length with $h_{A_2}(x)$. It should be noted that the selection of repeated element depends on the decision makers' risk preferences. The optimistic experts expect desirable outcomes and they may add the maximum element, while the pessimist experts expect unfavourable outcomes and they may add the minimum element. It is not out of mind that the final results may be different if one extends the shorter one by adding different elements. This is quite reasonable because the decision makers' risk preferences can directly influence the final decision [12, 19–29]. Hereafter, we extend the shorter one by adding the maximum value.

Definition 1.8 ([16]) Let A_1 and A_2 be two HFSs on X. Then, two kinds of ordering techniques for HFSs are presented as follows:

- The component-wise ordering of HFSs:

$$A_1 \leq A_2 \quad \text{if and only if} \quad \gamma_{A_1}^{\sigma(j)}(x_i) \leq \gamma_{A_2}^{\sigma(j)}(x_i), \quad 1 \leq i \leq n,\ 1 \leq j \leq l_{x_i}.$$

- The total ordering of HFSs:

$$A_1 \preceq A_2 \quad \text{if and only if} \quad Sc(A_1) \leq Sc(A_2),$$

in which $Sc(.)$ represents the score function of a HFS [54] given by

$$Sc(A_1) = \frac{1}{n}\sum_{i=1}^{n}\left(\frac{1}{l_{x_i}}\sum_{j=1}^{l_{x_i}}\gamma_{A_1}^{\sigma(j)}(x_i)\right). \tag{1.58}$$

Example 1.2 Let $X = \{x_1, x_2, x_3\}$ be the reference set, and

$$A_1 = \{\langle x_1, \{0.4\}\rangle, \langle x_2, \{0.3, 0.4\}\rangle, \langle x_3, \{0.3, 0.2, 0.5, 0.6\}\rangle\},$$
$$A_2 = \{\langle x_1, \{0.2, 0.5\}\rangle, \langle x_2, \{0.3\}\rangle, \langle x_3, \{0.3, 0.2\}\rangle\},$$

be two HFSs on X. Then, the HFSs A_1 and A_2 may be, respectively, represented by

$$A_1 = \{\langle x_1, \{0.4, 0.4\}\rangle, \langle x_2, \{0.3, 0.4\}\rangle, \langle x_3, \{0.2, 0.3, 0.5, 0.6\}\rangle\},$$
$$A_2 = \{\langle x_1, \{0.2, 0.5\}\rangle, \langle x_2, \{0.3, 0.3\}\rangle, \langle x_3, \{0.2, 0.3, 0.3, 0.3\}\rangle\}.$$

Using the first ordering technique presented in Definition 1.8, we easily find that the HFSs A_1 and A_2 are not component-wise comparable because $\gamma_{A_1}^{\sigma(1)}(x_1) \nleq \gamma_{A_2}^{\sigma(1)}(x_1)$ (therefore, $A_1 \nleq A_2$) and $\gamma_{A_2}^{\sigma(2)}(x_1) \nleq \gamma_{A_1}^{\sigma(2)}(x_1)$ (therefore, $A_2 \nleq A_1$).

Applying the second ordering technique represented in Definition 1.8 to the HFSs A_1 and A_2 gives rise to

$$Sc(A_1) = \frac{1}{3}\left(\frac{0.4+0.4}{2} + \frac{0.3+0.4}{2} + \frac{0.2+0.3+0.5+0.6}{4}\right) = 0.3833,$$
$$Sc(A_2) = \frac{1}{3}\left(\frac{0.2+0.5}{2} + \frac{0.3+0.3}{2} + \frac{0.2+0.3+0.3+0.3}{4}\right) = 0.3083.$$

This implies that A_2 is totally smaller than A_1 and denoted by $A_2 \preceq A_1$.

1.2 Hesitant Triangular Fuzzy Set

One of suitable tool for describing the uncertain membership degrees of an element is the form of fuzzy numbers. By taking on the concept of triangular fuzzy number into account, Yu [55] represented the concept of

(continued)

triangular fuzzy hesitant number, and further, defined the corresponding arithmetic operators. Shi et al. [46] defined Einstein correlated averaging operator to evaluate the wireless network security whose information is in the form of hesitant triangular fuzzy set. A development of hesitant triangular fuzzy set was given by Rashid and Husnine [43] who proposed trapezoidal valued hesitant fuzzy set. Peng [39] and Zhang et al. [60] developed some different multiple criteria decision making techniques whose criteria values are expressed by hesitant trapezoidal fuzzy elements. In a different manner of construction, Deli and Karaaslan [10] introduced the concept of generalized trapezoidal hesitant fuzzy set which is defined as a generalization of hesitant fuzzy sets and generalized fuzzy numbers.

The concept of hesitant triangular fuzzy set (HTFS) is an extension of HFS in which the membership degree of each element is stated by the use of triangular fuzzy numbers.

Definition 1.9 ([55]) A hesitant triangular fuzzy set (HTFS) $\underline{A}$ on X is generally denoted by

$$\underline{A} = \{\langle x, h_{\underline{A}}(x)\rangle \mid x \in X\}, \tag{1.59}$$

in which $h_{\underline{A}}(x)$ is referred to as the hesitant triangular fuzzy element (HTFE) and it represents the possible membership degrees of the element $x \in X$ to $\underline{A}$. In other words, the HTFS $\underline{A}$ is defined by the form of

$$\underline{A} = \left\{ \left\langle x, \bigcup_{(\gamma_{\underline{A}}^L, \gamma_{\underline{A}}^M, \gamma_{\underline{A}}^U) \in h_{\underline{A}}(x)} \{(\gamma_{\underline{A}}^L, \gamma_{\underline{A}}^M, \gamma_{\underline{A}}^U)\} \right\rangle \mid x \in X \right\}, \tag{1.60}$$

where

$$(\gamma_{\underline{A}}^L, \gamma_{\underline{A}}^M, \gamma_{\underline{A}}^U) = \begin{cases} \frac{t-\gamma_{\underline{A}}^L}{\gamma_{\underline{A}}^M-\gamma_{\underline{A}}^L} & \text{if} \quad \gamma_{\underline{A}}^L \le t \le \gamma_{\underline{A}}^M, \\ \frac{\gamma_{\underline{A}}^U-t}{\gamma_{\underline{A}}^U-\gamma_{\underline{A}}^M} & \text{if} \quad \gamma_{\underline{A}}^M \le t \le \gamma_{\underline{A}}^U. \end{cases}$$

It should be noticed that for any HTFE $h_{\underline{A}}(x)$, the condition $0 \le \gamma_{\underline{A}}^L, \gamma_{\underline{A}}^M, \gamma_{\underline{A}}^U \le 1$ must be satisfied.

Remark 1.2 Throughout this book, the set of all HTFSs on the reference set X is denoted by $\mathbb{HTFS}(X)$.

Example 1.3 Let $X = \{x_1, x_2\}$ be the reference set,
$h_{\underline{A}}(x_1) = \{(0.2, 0, 3, 0.45), (0.4, 0.6, 0.65), (0.5, 0.6, 0.7)\}$
and $h_{\underline{A}}(x_2) = \{(0.3, 0.5, 0.55), (0.4, 0.7, 0.8)\}$ be the HTFEs of x_i $(i = 1, 2)$ to a set $\underline{A}$, respectively. Then $\underline{A}$ can be considered as an HTFS, i.e.,

$$\underline{A} = \{\langle x_1, \{(0.2, 0, 3, 0.45), (0.4, 0.6, 0.65), (0.5, 0.6, 0.7)\}\rangle, \\ \langle x_2, \{(0.3, 0.5, 0.55), (0.4, 0.7, 0.8)\}\rangle\}.$$

Definition 1.10 ([9]) A hesitant trapezoidal fuzzy set (HTFS) $\underline{A}$ on X is generally denoted by

$$\underline{A} = \{\langle x, h_{\underline{A}}(x)\rangle \mid x \in X\}, \tag{1.61}$$

where $h_{\underline{A}}(x)$ is referred to as the hesitant trapezoidal fuzzy element (HTFE) and it represents the possible membership degrees of the element $x \in X$ to $\underline{A}$. In other words, the HTFS $\underline{A}$ can be denoted by

$$\underline{A} = \left\{\left\langle x, \bigcup_{(\gamma_{\underline{A}}^{L}, \gamma_{\underline{A}}^{M_L}, \gamma_{\underline{A}}^{M_U}, \gamma_{\underline{A}}^{U}) \in h_{\underline{A}}(x)} \{(\gamma_{\underline{A}}^{L}, \gamma_{\underline{A}}^{M_L}, \gamma_{\underline{A}}^{M_U}, \gamma_{\underline{A}}^{U})\}\right\rangle \mid x \in X\right\},$$

where

$$(\gamma_{\underline{A}}^{L}, \gamma_{\underline{A}}^{M_L}, \gamma_{\underline{A}}^{M_U}, \gamma_{\underline{A}}^{U}) = \begin{cases} \frac{t-\gamma_{\underline{A}}^{L}}{\gamma_{\underline{A}}^{M_L}-\gamma_{\underline{A}}^{L}} & \text{if } \gamma_{\underline{A}}^{L} \le t \le \gamma_{\underline{A}}^{M_L}, \\ 1 & \text{if } \gamma_{\underline{A}}^{M_L} \le t \le \gamma_{\underline{A}}^{M_U}, \\ \frac{\gamma_{\underline{A}}^{U}-t}{\gamma_{\underline{A}}^{U}-\gamma_{\underline{A}}^{M_U}} & \text{if } \gamma_{\underline{A}}^{M_U} \le t \le \gamma_{\underline{A}}^{U}. \end{cases}$$

It should be noticed that for any HTFE $h_{\underline{A}}(x)$, the condition $0 \le \gamma_{\underline{A}}^{L}, \gamma_{\underline{A}}^{M_L}, \gamma_{\underline{A}}^{M_U}, \gamma_{\underline{A}}^{U} \le 1$ must be satisfied.

Remark 1.3 Throughout this book, the set of all HTFSs on the reference set X is denoted by $\mathbb{HTFS}(X)$.

Definition 1.11 ([10]) A generalized trapezoidal hesitant fuzzy set (GTHFS) $\underline{A}$ on X is denoted by

$$\underline{A} = \{\langle x, h_{\underline{A}}(x)\rangle \mid x \in X\}, \tag{1.62}$$

where $h_{\underline{A}}(x)$ is referred to as the generalized trapezoidal hesitant fuzzy element (GTHFE), and it is in fact a set of some different trapezoidal fuzzy values together with a hesitant fuzzy set in [0, 1]. In other words, the GTHFS $\underline{A}$ can be presented by

$$\underline{A} = \left\{ \left\langle x, \bigcup_{\langle(\gamma_{\underline{A}}^{L}, \gamma_{\underline{A}}^{M_L}, \gamma_{\underline{A}}^{M_U}, \gamma_{\underline{A}}^{U}), \bigcup_{\gamma_{\underline{A}} \in [0,1]}\{\gamma_{\underline{A}}\}\rangle \in h_{\underline{A}}(x)} \{\langle(\gamma_{\underline{A}}^{L}, \gamma_{\underline{A}}^{M_L}, \gamma_{\underline{A}}^{M_U}, \gamma_{\underline{A}}^{U}), \right.\right.$$

$$\left.\left. \bigcup_{\gamma_{\underline{A}} \in [0,1]} \{\gamma_{\underline{A}}\}\rangle\} \right\rangle \mid x \in X \right\}. \tag{1.63}$$

where

$$(\gamma_{\underline{A}}^{L}, \gamma_{\underline{A}}^{M_L}, \gamma_{\underline{A}}^{M_U}, \gamma_{\underline{A}}^{U}) = \begin{cases} \frac{t-\gamma_{\underline{A}}^{L}}{\gamma_{\underline{A}}^{M_L}-\gamma_{\underline{A}}^{L}} & \text{if} \quad \gamma_{\underline{A}}^{L} \leq t \leq \gamma_{\underline{A}}^{M_L}, \\ 1 & \text{if} \quad \gamma_{\underline{A}}^{M_L} \leq t \leq \gamma_{\underline{A}}^{M_U}, \\ \frac{\gamma_{\underline{A}}^{U}-t}{\gamma_{\underline{A}}^{U}-\gamma_{\underline{A}}^{M_U}} & \text{if} \quad \gamma_{\underline{A}}^{M_U} \leq t \leq \gamma_{\underline{A}}^{U}, \end{cases}$$

and $\bigcup_{\gamma_{\underline{A}} \in [0,1]}\{\gamma_{\underline{A}}\}$ is a HFE.

Remark 1.4 Throughout this book, the set of all GTHFSs on the reference set X is denoted by $\mathbb{GTHFS}(X)$.

Example 1.4 Let $X = \{x_1, x_2\}$ be the reference set,
$h_{\underline{A}}(x_1) = \{\langle(0.2, 0, 3, 0.45, 0.5), \{0.4, 0.6, 0.65\}\rangle\}$ and
$h_{\underline{A}}(x_2) = \{\langle(0.3, 0.5, 0.55, 0.6), \{0.4, 0.7, 0.8\}\rangle\}$ be the GTHFEs of x_i $(i = 1, 2)$ to a set $\underline{A}$, respectively. Then $\underline{A}$ can be considered as an GTHFS, i.e.,

$$\underline{A} = \{\langle x_1, \langle(0.2, 0, 3, 0.45, 0.5), \{0.4, 0.6, 0.65\}\rangle\rangle,$$
$$\langle x_2, \langle(0.3, 0.5, 0.55, 0.6), \{0.4, 0.7, 0.8\}\rangle\rangle\}.$$

Given three GTHFEs represented by $h_{\underline{A}}$, $h_{\underline{A}_1}$, and $h_{\underline{A}_2}$, two fundamental arithmetic operations on the GTHFEs, which are also GTHFEs, are described by the following (see, e.g., [10]):

$$h_{\underline{A}_1} \oplus h_{\underline{A}_2}$$
$$= \bigcup_{\substack{\langle(\gamma_{\underline{A}_1}^{L}, \gamma_{\underline{A}_1}^{M_L}, \gamma_{\underline{A}_1}^{M_U}, \gamma_{\underline{A}_1}^{U}), \bigcup_{\gamma_{\underline{A}_1} \in [0,1]}\{\gamma_{\underline{A}_1}\}\rangle \in h_{\underline{A}_1}(x)\rangle \\ \langle(\gamma_{\underline{A}_2}^{L}, \gamma_{\underline{A}_2}^{M_L}, \gamma_{\underline{A}_2}^{M_U}, \gamma_{\underline{A}_2}^{U}), \bigcup_{\gamma_{\underline{A}_2} \in [0,1]}\{\gamma_{\underline{A}_2}\}\rangle \in h_{\underline{A}_2}(x)\rangle}}$$

$$\times \left\{ \left\langle (\gamma_{\underline{A}_1}^{L} + \gamma_{\underline{A}_2}^{L}, \gamma_{\underline{A}_1}^{M_L} + \gamma_{\underline{A}_2}^{M_L}, \gamma_{\underline{A}_1}^{M_U} + \gamma_{\underline{A}_2}^{M_U}, \gamma_{\underline{A}_1}^{U} + \gamma_{\underline{A}_2}^{U}), \bigcup_{\gamma_{\underline{A}_1}, \gamma_{\underline{A}_2} \in [0,1]} \{\gamma_{\underline{A}_1} + \gamma_{\underline{A}_2} - \gamma_{\underline{A}_1} \times \gamma_{\underline{A}_2}\} \right\rangle \right\}; \tag{1.64}$$

$$h_{\underline{A}_1} \otimes h_{\underline{A}_2} = \bigcup_{\substack{\langle(\gamma_{\underline{A}_1}^{L}, \gamma_{\underline{A}_1}^{M_L}, \gamma_{\underline{A}_1}^{M_U}, \gamma_{\underline{A}_1}^{U}), \bigcup_{\gamma_{\underline{A}_1} \in [0,1]}\{\gamma_{\underline{A}_1}\}\rangle \in h_{\underline{A}_1}(x) \\ \langle(\gamma_{\underline{A}_2}^{L}, \gamma_{\underline{A}_2}^{M_L}, \gamma_{\underline{A}_2}^{M_U}, \gamma_{\underline{A}_2}^{U}), \bigcup_{\gamma_{\underline{A}_2} \in [0,1]}\{\gamma_{\underline{A}_2}\}\rangle \in h_{\underline{A}_2}(x)}}$$

$$\times \left\{ \left\langle (\gamma_{\underline{A}_1}^{L} \times \gamma_{\underline{A}_2}^{L}, \gamma_{\underline{A}_1}^{M_L} \times \gamma_{\underline{A}_2}^{M_L}, \gamma_{\underline{A}_1}^{M_U} \times \gamma_{\underline{A}_2}^{M_U}, \gamma_{\underline{A}_1}^{U} \times \gamma_{\underline{A}_2}^{U}), \bigcup_{\gamma_{\underline{A}_1}, \gamma_{\underline{A}_2} \in [0,1]} \{\gamma_{\underline{A}_1} \times \gamma_{\underline{A}_2}\} \right\rangle \right\}; \tag{1.65}$$

$$h_{\underline{A}}^{\lambda} = \bigcup_{\langle(\gamma_{\underline{A}}^{L}, \gamma_{\underline{A}}^{M_L}, \gamma_{\underline{A}}^{M_U}, \gamma_{\underline{A}}^{U}), \bigcup_{\gamma_{\underline{A}} \in [0,1]}\{\gamma_{\underline{A}}\}\rangle \in h_{\underline{A}}(x)} \left\{ \left\langle (\gamma_{\underline{A}}^{L\lambda}, \gamma_{\underline{A}}^{M_L\lambda}, \gamma_{\underline{A}}^{M_U\lambda}, \gamma_{\underline{A}}^{U\lambda}), \bigcup_{\gamma_{\underline{A}} \in [0,1]} \{(\gamma_{\underline{A}})^{\lambda}\} \right\rangle \right\}; \tag{1.66}$$

$$\lambda h_{\underline{A}} = \bigcup_{\langle(\gamma_{\underline{A}}^{L}, \gamma_{\underline{A}}^{M_L}, \gamma_{\underline{A}}^{M_U}, \gamma_{\underline{A}}^{U}), \bigcup_{\gamma_{\underline{A}} \in [0,1]}\{\gamma_{\underline{A}}\}\rangle \in h_{\underline{A}}(x)} \left\{ \left\langle (\lambda\gamma_{\underline{A}}^{L}, \lambda\gamma_{\underline{A}}^{M_L}, \lambda\gamma_{\underline{A}}^{M_U}, \lambda\gamma_{\underline{A}}^{U}), \bigcup_{\gamma_{\underline{A}} \in [0,1]} \{1 - (1 - \gamma_{\underline{A}})^{\lambda}\} \right\rangle \right\}, \tag{1.67}$$

where $\lambda > 0$.

1.3 Interval-Valued Hesitant Fuzzy Set

In many practical applications, the uncertainty is usually specified between some different values, and determination of crisp membership degrees is not easy task. This impulsed Chen et al. [8] to define a generalization of hesitant fuzzy set as the concept of interval-valued hesitant fuzzy set which dominates the barrier and helps a decision maker to assign some membership degrees for an object under a set to have several interval values [41]. The theory of interval-valued hesitant fuzzy set has attracted more attention in decision making fields [34, 51], and in group decision making processing [8]. Li and Peng [34] introduced the interval-valued hesitant fuzzy Hamacher synergetic weighted aggregation operators; Wei et al. [51] proposed a number of hesitant interval-valued fuzzy aggregation operators; and Chen et al. [8] presented some interval-valued hesitant preference relations.

The concept of interval-valued hesitant fuzzy set as a generalization of HFS is similar to that encountered in intuitionistic fuzzy environments, where the concept of intuitionistic fuzzy set has been extended to that of interval-valued intuitionistic fuzzy set.

Definition 1.12 ([8]) An interval-valued hesitant fuzzy set (IVHFS) $\widetilde{A}$ on X is defined as:

$$\widetilde{A} = \{\langle x, h_{\widetilde{A}}(x)\rangle \mid x \in X\}, \tag{1.68}$$

where $h_{\widetilde{A}}(x)$ is referred to as the interval-valued hesitant fuzzy element (IVHFE) and it is in fact a set of some different interval values in $[0, 1]$. Further, $h_{\widetilde{A}}(x)$ represents the possible membership degrees of the element $x \in X$ to $\widetilde{A}$. In this regards, the IVHFS $\widetilde{A}$ can be denoted by

$$\widetilde{A} = \left\{\left\langle x, \bigcup_{[\gamma^L_{\widetilde{A}}, \gamma^U_{\widetilde{A}}] \in h_{\widetilde{A}}(x)} \{[\gamma^L_{\widetilde{A}}, \gamma^U_{\widetilde{A}}]\}\right\rangle \mid x \in X\right\}. \tag{1.69}$$

Remark 1.5 Throughout this book, the set of all IVHFSs on the reference set X is denoted by $\mathbb{IVHFS}(X)$.

Example 1.5 Let $X = \{x_1, x_2\}$ be the reference set, $h_{\widetilde{A}}(x_1) = \{[0.2, 0, 3], [0.4, 0.6], [0.5, 0.6]\}$ and $h_{\widetilde{A}}(x_2) = \{[0.3, 0.5], [0.4, 0.7]\}$ be the IVHFEs of x_i $(i = 1, 2)$ to a set $\widetilde{A}$, respectively. Then $\widetilde{A}$ can be considered as an IVHFS, i.e.,

$$\widetilde{A} = \{\langle x_1, \{[0.2, 0, 3], [0.4, 0.6], [0.5, 0.6]\}\rangle, \langle x_2, \{[0.3, 0.5], [0.4, 0.7]\}\rangle\}.$$

Given three IVHFEs represented by $h_{\widetilde{A}}$, $h_{\widetilde{A}_1}$, and $h_{\widetilde{A}_2}$, some set and arithmetic operations on the IVHFEs, which are also IVHFEs, can be described as follows (see, e.g., [8]):

$$h_{\widetilde{A}}^c = \bigcup_{[\gamma_{\widetilde{A}}^L, \gamma_{\widetilde{A}}^U] \in h_{\widetilde{A}}} \{[1 - \gamma_{\widetilde{A}}^U, 1 - \gamma_{\widetilde{A}}^L]\}; \tag{1.70}$$

$$h_{\widetilde{A}_1} \cup h_{\widetilde{A}_2} = \bigcup_{[\gamma_{\widetilde{A}_1}^L, \gamma_{\widetilde{A}_1}^U] \in h_{\widetilde{A}_1}, [\gamma_{\widetilde{A}_2}^L, \gamma_{\widetilde{A}_2}^U] \in h_{\widetilde{A}_2}} \{[\max\{\gamma_{\widetilde{A}_1}^L, \gamma_{\widetilde{A}_2}^L\}, \max\{\gamma_{\widetilde{A}_1}^U, \gamma_{\widetilde{A}_2}^U\}]\}; \tag{1.71}$$

$$h_{\widetilde{A}_1} \cap h_{\widetilde{A}_2} = \bigcup_{[\gamma_{\widetilde{A}_1}^L, \gamma_{\widetilde{A}_1}^U] \in h_{\widetilde{A}_1}, [\gamma_{\widetilde{A}_2}^L, \gamma_{\widetilde{A}_2}^U] \in h_{\widetilde{A}_2}} \{[\min\{\gamma_{\widetilde{A}_1}^L, \gamma_{\widetilde{A}_2}^L\}, \min\{\gamma_{\widetilde{A}_1}^U, \gamma_{\widetilde{A}_2}^U\}]\}; \tag{1.72}$$

$$h_{\widetilde{A}_1} \oplus h_{\widetilde{A}_2} = \bigcup_{[\gamma_{\widetilde{A}_1}^L, \gamma_{\widetilde{A}_1}^U] \in h_{\widetilde{A}_1}, [\gamma_{\widetilde{A}_2}^L, \gamma_{\widetilde{A}_2}^U] \in h_{\widetilde{A}_2}} \{[\gamma_{\widetilde{A}_1}^L + \gamma_{\widetilde{A}_2}^L - \gamma_{\widetilde{A}_1}^L \gamma_{\widetilde{A}_2}^L, \gamma_{\widetilde{A}_1}^U + \gamma_{\widetilde{A}_2}^U - \gamma_{\widetilde{A}_1}^U \gamma_{\widetilde{A}_2}^U]\}; \tag{1.73}$$

$$h_{\widetilde{A}_1} \otimes h_{\widetilde{A}_2} = \bigcup_{[\gamma_{\widetilde{A}_1}^L, \gamma_{\widetilde{A}_1}^U] \in h_{\widetilde{A}_1}, [\gamma_{\widetilde{A}_2}^L, \gamma_{\widetilde{A}_2}^U] \in h_{\widetilde{A}_2}} \{[\gamma_{\widetilde{A}_1}^L \gamma_{\widetilde{A}_2}^L, \gamma_{\widetilde{A}_1}^U \gamma_{\widetilde{A}_2}^U]\}; \tag{1.74}$$

$$h_{\widetilde{A}}^\lambda = \bigcup_{[\gamma_{\widetilde{A}}^L, \gamma_{\widetilde{A}}^U] \in h_{\widetilde{A}}} \{[(\gamma_{\widetilde{A}}^L)^\lambda, (\gamma_{\widetilde{A}}^U)^\lambda]\}, \quad \lambda > 0; \tag{1.75}$$

$$\lambda h_{\widetilde{A}} = \bigcup_{[\gamma_{\widetilde{A}}^L, \gamma_{\widetilde{A}}^U] \in h_{\widetilde{A}}} \{[1 - (1 - \gamma_{\widetilde{A}}^L)^\lambda, 1 - (1 - \gamma_{\widetilde{A}}^U)^\lambda]\}, \quad \lambda > 0. \tag{1.76}$$

Continuing in the same vein, we are easily able to obtain the product and sum operations of Einstein, Hamacher, and Frank for IVHFEs whose Algebraic sum and Algebraic product operations have already presented by Eqs. (1.73) and (1.74), above.

Given two IVHFEs represented by $h_{\widetilde{A}_1} = \bigcup_{[\gamma^L_{\widetilde{A}_1},\gamma^U_{\widetilde{A}_1}]\in h_{\widetilde{A}_1}}\{[\gamma^L_{\widetilde{A}_1},\gamma^U_{\widetilde{A}_1}]\}$ and $h_{\widetilde{A}_2} = \bigcup_{[\gamma^L_{\widetilde{A}_2},\gamma^U_{\widetilde{A}_2}]\in h_{\widetilde{A}_2}}\{[\gamma^L_{\widetilde{A}_2},\gamma^U_{\widetilde{A}_2}]\}$, several division and subtraction operations on the IVHFEs being also IVHFEs can be described as follows:

$$h_{\widetilde{A}_1}\ominus h_{\widetilde{A}_2} = \bigcup_{[\gamma^L_{\widetilde{A}_1},\gamma^U_{\widetilde{A}_1}]\in h_{\widetilde{A}_1},[\gamma^L_{\widetilde{A}_2},\gamma^U_{\widetilde{A}_2}]\in h_{\widetilde{A}_2}}\left[\max\left\{0,\frac{\gamma^L_{\widetilde{A}_1}-\gamma^L_{\widetilde{A}_2}}{1-\gamma^L_{\widetilde{A}_2}}\right\},\max\left\{0,\frac{\gamma^U_{\widetilde{A}_1}-\gamma^U_{\widetilde{A}_2}}{1-\gamma^U_{\widetilde{A}_2}}\right\}\right], \tag{1.77}$$

$$h_{\widetilde{A}_1}\oslash h_{\widetilde{A}_2} = \bigcup_{[\gamma^L_{\widetilde{A}_1},\gamma^U_{\widetilde{A}_1}]\in h_{\widetilde{A}_1},[\gamma^L_{\widetilde{A}_2},\gamma^U_{\widetilde{A}_2}]\in h_{\widetilde{A}_2}}\left[\min\left\{1,\frac{\gamma^L_{\widetilde{A}_1}}{\gamma^L_{\widetilde{A}_2}}\right\},\min\left\{1,\frac{\gamma^U_{\widetilde{A}_1}}{\gamma^U_{\widetilde{A}_2}}\right\}\right], \tag{1.78}$$

where the right hand side operations can be taken as Einstein, Hamacher, or Frank division and subtraction operations proposed in Definition 1.7.

1.4 Extended Hesitant Fuzzy Set

As an extension of the hesitant fuzzy set concept, Zhu and Xu [62] proposed the concept of extended hesitant fuzzy set as a tool to avoid loss of information. In fact, such a consideration enables us to present information represented by the decision makers using possible value-groups. However, this definition of extended hesitant fuzzy set suffers from some drawbacks which are mentioned by a fresh study conducted by Farhadinia and Herrera-Viedma [25]. They revisited the concept of extended hesitant fuzzy set proposed by Zhu and Xu using the Cartesian product of hesitant fuzzy sets. Farhadinia and Xu [30] have recently dealt with a new aspect of emergency event that used instead of usual aggregation procedure, a new fusion technique based on the modified version of extended hesitant fuzzy set.

Definition 1.13 ([62]) Suppose that X is the reference set, and $h_k(x) = \{\gamma_k^{\delta(i)}(x) \mid i = 1,\ldots,l_{h_k}\}$ for $k = 1,..,m_x$ denote a family of m_x hesitant fuzzy elements defined for a given $x \in X$. Then, the version of Zhu and Xu's [62] extended hesitant fuzzy set (Z-EHFS) is defined as

$$
\begin{aligned}
\grave{A} &= \{\langle x, h_{\grave{A}}(x)\rangle \mid x \in X\} \\
&= \{\langle x, h_1(x) \times h_2(x) \times \ldots \times h_{m_x}(x)\rangle \mid x \in X\} \\
&= \left\{\left\langle x, \bigcup_{(\gamma_1(x),\ldots,\gamma_m(x))\in h_1(x)\times h_2(x)\times\ldots\times h_{m_x}(x)} \{(\gamma_1(x), \ldots, \gamma_m(x))\}\right\rangle \mid x \in X\right\}.
\end{aligned}
\tag{1.79}
$$

Before any progress can be made in defining the EHFS, it is necessary to take this point into consideration that each element of EHFS, called hereafter as the extended hesitant fuzzy element (EHFE), is a set of n-tuples which indicates the opinion of n number of decision makers, simultaneously.

To say more precisely, Farhadinia and Herrera-Viedma [25] considered a situation in which an EHFE is used to indicate the opinion of $n = 3$ decision makers simultaneously, meanwhile the number of HFEs is 4. In such a situation, each EHFE contains of the 3-tuple elements, meanwhile the Cartesian product of 4 HFEs is a set of 4-tuple elements. This clearly shows that an EHFE including 3-tuple elements is different from the Cartesian product of 4 HFEs. Hence, an element of EHFE does not generally play the role of the element included in the Cartesian product of HFEs as defined in Zhu and Xu [62].

Farhadinia and Herrera-Viedma [25] re-defined the EHFS as the following form.

Definition 1.14 ([25]) Suppose that X is the reference set. Then, an extended hesitant fuzzy set (EHFS) is characterized by the following mathematical symbol

$$
\begin{aligned}
\grave{A} &= \{\langle x, h_{\grave{A}}(x)\rangle \mid x \in X\} \\
&= \left\{\left\langle x, \bigcup_{(\gamma_1(x),\ldots,\gamma_m(x))\in h_{\grave{A}}(x)} \{(\gamma_1(x), \ldots, \gamma_m(x))\}\right\rangle \mid x \in X\right\},
\end{aligned}
\tag{1.80}
$$

where $h_{\grave{A}}(x)$ stands for an extended hesitant fuzzy element (EHFE).

Example 1.6 Suppose that $X = \{x_1, x_2\}$ is the reference set, and $h_{\grave{A}_1}(x) = \{(0.5, 0.4, 0.3), (0.5, 0.3, 0.2)\}$ and $h_{\grave{A}_2}(x) = \{(0.5, 0.2, 0.2)\}$ are two EHFEs on X. Then, the EHFS $\grave{A}$ is characterized by

$$
\begin{aligned}
\grave{A} &= \{\langle x_1, h_{\grave{A}_1}(x)\rangle, \langle x_2, h_{\grave{A}_2}(x)\rangle\} \\
&= \{\langle x_1, \{(0.5, 0.4, 0.3), (0.5, 0.3, 0.2)\}\rangle, \langle x_2, \{(0.5, 0.2, 0.2)\}\rangle\}.
\end{aligned}
$$

Here, we are going to introduce some operational laws on the EHFEs that help to aggregate the extended hesitant fuzzy information.

Considering the notation $\delta(*)$ as the $*$-th element of the EHFE $h_{\grave{A}}$, we define the following operations on the EHFEs $h_{\grave{A}_1} = \{\gamma_{\grave{A}_1}^{\delta(t)} := (\gamma_1^{1,\delta(t)}, \ldots, \gamma_m^{1,\delta(t)}) \mid t = 1, \ldots, l_{h_1}\}$ and $h_{\grave{A}_2} = \{\gamma_{\grave{A}_2}^{\delta(r)} := (\gamma_1^{2,\delta(r)}, \ldots, \gamma_m^{2,\delta(r)}) \mid r = 1, \ldots, l_{h_2}\}$ in the forms of

$$\begin{aligned} h_{\grave{A}_1} \oplus h_{\grave{A}_2} &= \bigcup_{\gamma_{\grave{A}_1}^{\delta(t)} \in h_{\grave{A}_1}, \gamma_{\grave{A}_2}^{\delta(r)} \in h_{\grave{A}_2}} \{\gamma_{\grave{A}_1}^{\delta(t)} + \gamma_{\grave{A}_2}^{\delta(r)} - \gamma_{\grave{A}_1}^{\delta(t)} \gamma_{\grave{A}_2}^{\delta(r)}\} \\ &= \bigcup_{(\gamma_1^{1,\delta(t)}, \ldots, \gamma_m^{1,\delta(t)}) \in h_{\grave{A}_1}, (\gamma_1^{2,\delta(r)}, \ldots, \gamma_m^{2,\delta(r)}) \in h_{\grave{A}_2}} \{(\gamma_1^{1,\delta(t)} + \gamma_1^{2,\delta(r)} - \gamma_1^{1,\delta(t)} \gamma_1^{2,\delta(r)}, \ldots, \\ &\quad \gamma_m^{1,\delta(t)} + \gamma_m^{2,\delta(r)} - \gamma_m^{1,\delta(t)} \gamma_m^{2,\delta(r)})\}; \end{aligned} \tag{1.81}$$

$$\begin{aligned} h_{\grave{A}_1} \otimes h_{\grave{A}_2} &= \bigcup_{\gamma_{\grave{A}_1}^{\delta(t)} \in h_{\grave{A}_1}, \gamma_{\grave{A}_2}^{\delta(r)} \in h_{\grave{A}_2}} \{\gamma_{\grave{A}_1}^{\delta(t)} \gamma_{\grave{A}_2}^{\delta(r)}\} \\ &= \bigcup_{(\gamma_1^{1,\delta(t)}, \ldots, \gamma_m^{1,\delta(t)}) \in h_{\grave{A}_1}, (\gamma_1^{2,\delta(r)}, \ldots, \gamma_m^{2,\delta(r)}) \in h_{\grave{A}_2}} \{(\gamma_1^{1,\delta(t)} \gamma_1^{2,\delta(r)}, \ldots, \gamma_m^{1,\delta(t)} \gamma_m^{2,\delta(r)})\}; \end{aligned} \tag{1.82}$$

$$\begin{aligned} \lambda h_{\grave{A}_1} &= \bigcup_{\gamma_{\grave{A}_1}^{\delta(t)} \in h_{\grave{A}_1}} \{1 - (1 - \gamma_{\grave{A}_1}^{\delta(t)})^{\lambda}\} \\ &= \bigcup_{(\gamma_1^{1,\delta(t)}, \ldots, \gamma_m^{1,\delta(t)}) \in h_{\grave{A}_1}} \{(1 - (1 - \gamma_1^{1,\delta(t)})^{\lambda}, \ldots, 1 - (1 - \gamma_m^{1,\delta(t)})^{\lambda})\}, \quad \lambda > 0; \end{aligned} \tag{1.83}$$

$$\begin{aligned} h_{\grave{A}_1}^{\lambda} &= \bigcup_{\gamma_{\grave{A}_1}^{\delta(t)} \in h_{\grave{A}_1}} \{(\gamma_{\grave{A}_1}^{\delta(t)})^{\lambda}\} \\ &= \bigcup_{(\gamma_1^{1,\delta(t)}, \ldots, \gamma_m^{1,\delta(t)}) \in h_{\grave{A}_1}} \{([\gamma_1^{1,\delta(t)}]^{\lambda}, \ldots, [\gamma_m^{1,\delta(t)}]^{\lambda})\}, \quad \lambda > 0. \end{aligned} \tag{1.84}$$

1.5 Higher Order Hesitant Fuzzy Set

The generalization of hesitant fuzzy set, known as the generalized hesitant fuzzy set [40], has its inherent drawbacks, because it expresses the membership degrees of an element to a given set only by crisp numbers or intuitionistic fuzzy sets. In many practical decision making problems, the information provided by a decision maker might often be described by fuzzy sets (instead of crisp numbers) or other fuzzy set extensions (instead of intuitionistic fuzzy sets). Therefore, it is difficult for the decision makers to

(continued)

provide exact crisp values or just intuitionistic fuzzy sets for the membership degrees. This difficulty can be avoided using a higher order of hesitant fuzzy set introduced by Farhadinia [14] for the membership degrees. The higher order hesitant fuzzy set is more fit for the case when the decision makers have a hesitation among several possible memberships with uncertainties.

Before going more deeply into the matter, it might be well to introduce the concept which was first presented by Farhadinia [14].

We suppose that X to be a reference set. Farhadinia [14] defined a generalized type of fuzzy set (G-Type FS) on X in the form of

$$A = \{\langle x, h_A(x)\rangle \mid x \in X\}, \tag{1.85}$$

where

$$h_A : X \to \psi([0, 1]).$$

Here, $\psi([0, 1])$ denotes a family of crisp or fuzzy sets that can be defined with in the universal set $[0, 1]$.

It is noteworthy that most of the existing extensions of ordinary fuzzy set are special cases of G-Type FS. For instance, by taking G-Type FS A given by (1.85) into consideration, we conclude that (see [33])

- If $\psi([0, 1]) = [0, 1]$, then the G-Type FS A reduces to an ordinary fuzzy set;
- If $\psi([0, 1]) = \varepsilon([0, 1])$ denoting the set of all closed intervals, then the G-Type FS A reduces to an interval-valued fuzzy set;
- If $\psi([0, 1]) = F([0, 1])$ denoting the set of all ordinary fuzzy sets, then the G-Type FS A reduces to a type 2 fuzzy set;
- If $\psi([0, 1]) = L$ denoting a partially ordered Lattice, then the G-Type FS A reduces to a lattice fuzzy set.

As can be seen from definition of HFS, it expresses the membership degrees of an element to a given set only by several real numbers between 0 and 1, while in many real-world situations assigning exact values to the membership degrees does not describe properly the imprecise or uncertain decision information. Thus, it seems to be not easy for the decision makers to rely on HFSs for expressing uncertainty of an element.

To overcome the difficulty associated with expressing uncertainty of an element to a given set, Farhadinia [14] introduced the concept of higher order hesitant fuzzy set (HOHFS). This makes the membership degrees of an element for a given set to be expressed by several possible G-Type FSs.

Definition 1.15 ([14]) Let X be a reference set. A higher order hesitant fuzzy set (HOHFS) on X is defined in terms of a function that when applied to X returns a set of G-Type FSs. A HOHFS is denoted by

$$\breve{A} = \{\langle x, h_{\breve{A}}(x)\rangle \mid x \in X\}, \tag{1.86}$$

in which $h_{\breve{A}}(x)$ is referred to as the higher order hesitant fuzzy element (HOHFE), and it is a set of some G-Type FSs denoting the possible membership degree of the element $x \in X$ to the set $\breve{A}$. In this regards, the HOHFS $\breve{A}$ is also represented as:

$$\breve{A} = \left\{\left\langle x, \bigcup_{\gamma_{\breve{A}} \in h_{\breve{A}}} \{\gamma_{\breve{A}}\}\right\rangle \mid x \in X\right\},$$

where all $\gamma_{\breve{A}}(x)$ are G-Type FSs on X.

Remark 1.6 Throughout this book, the set of all HOHFSs on the reference set X is denoted by $\mathbb{HOHFS}(X)$.

Example 1.7 If $X = \{x_1, x_2, x_3\}$ is the reference set, then

$$h_{\breve{A}}(x_1) = \{(0.2, 0.4), (0.5, 0.3)\},$$
$$h_{\breve{A}}(x_2) = \{(0.3, 0.4)\},$$
$$h_{\breve{A}}(x_3) = \{(0.3, 0.2), (0.1, 0.3), (0.5, 0.4)\}$$

are the HOHFEs of x_i $(i = 1, 2, 3)$ to the set $\breve{A}$ where the G-Type FSs $\gamma_{\breve{A}}^{\sigma(k)}(x_i) = (\mu_{ki}, \nu_{ki})$ are intuitionistic fuzzy sets such that $0 \leq \mu_{ki}, \nu_{ki} \leq 1$ and $0 \leq \mu_{ki}+\nu_{ki} \leq 1$ for $k = 1, 2, \ldots, l_{x_i}$ $(i = 1, 2, \ldots, |X| = 3)$. Then $\breve{A}$ can be considered as a HOHFS, i.e.,

$$\breve{A} = \{\langle x_1, \{(0.2, 0.4), (0.5, 0.3)\}\rangle, \langle x_2, \{(0.3, 0.4)\}\rangle,$$
$$\langle x_3, \{(0.3, 0.2), (0.1, 0.3), (0.5, 0.4)\}\rangle\}.$$

In view of Definition 1.15 and the latter review of some fuzzy set extensions, it is easily deduced that each HOHFS becomes a type 2 fuzzy set if all its G-Type FSs are the same. That is, if $h_{\breve{A}}(x) := \gamma_{\breve{A}}^{\sigma(1)}(x) = \ldots = \gamma_{\breve{A}}^{\sigma(l_x)}(x)$ for any $x \in X$, then the HOHFS $\breve{A} = \{\langle x, h_{\breve{A}}(x)\rangle \mid x \in X\}$ reduces to a type 2 fuzzy set.

It is noteworthy to mention again that the notions of interval-valued hesitant fuzzy set [13] given in the next sections and the interval type-2 fuzzy set [37] both are special cases of HOHFSs. A HOHFS $\breve{A} = \{\langle x, h_{\breve{A}}(x)\rangle \mid x \in X\}$ reduces to an interval-valued hesitant fuzzy set, when all G-Type FSs $\gamma_{\breve{A}}^{\sigma(t)}(x)$ $(t = 1, \ldots, l_x)$ for any $x \in X$ are considered as closed intervals of real numbers in [0, 1]. Furthermore, an interval-valued hesitant fuzzy set $\breve{A} = \{\langle x, \bigcup_{\gamma_{\breve{A}}^{\sigma(t)} \in h_{\breve{A}}} \{\gamma_{\breve{A}}^{\sigma(t)}\}\rangle \mid x \in X\}$ reduces to an interval-valued type 2 fuzzy set, when all intervals satisfy $\gamma_{\breve{A}}^{\sigma(1)}(x) := [\gamma_{\breve{A}}^{\sigma(1),L}(x), \gamma_{\breve{A}}^{\sigma(1),U}(x)] = \ldots = \gamma_{\breve{A}}^{\sigma(l_x)}(x) := [\gamma_{\breve{A}}^{\sigma(l_x),L}(x), \gamma_{\breve{A}}^{\sigma(l_x),U}(x)]$ for any $x \in X$.

Qian et al. [40] extended HFSs by the use of intuitionistic fuzzy set concept and referred to that as the generalized hesitant fuzzy set (G-HFS). They stated that fuzzy sets, intuitionistic fuzzy sets, and HFSs are special cases of G-HFSs. Obviously, a G-HFS $\breve{A} = \{\langle x, h_{\breve{A}}(x)\rangle \mid x \in X\}$ is also a special case of HOHFS where all G-Type FSs $\gamma_{\breve{A}}^{\sigma(1)}(x), \ldots, \gamma_{\breve{A}}^{\sigma(l_x)}(x)$ for any $x \in X$ are considered as intuitionistic fuzzy sets.

1.6 Dual Hesitant Fuzzy Set

Dual hesitant fuzzy set is a representation which seeks to combine intuitionist fuzzy concept and hesitant fuzzy concept by uniting the advantages of these concepts. Indeed, Zhu et al. [63] introduced the concept of dual hesitant fuzzy set which can encompass fuzzy sets, intuitionistic fuzzy sets, hesitant fuzzy sets, and fuzzy multisets as special cases. It is seen that dual hesitant fuzzy sets can better deal with the situations that permit the membership and the non-membership of an element to a given set having a few different values. Such situations often arise in group decision making problems, such as, the study deals with a number of dual hesitant fuzzy information aggregation operators including mean and geometric forms of dual hesitant fuzzy Heronian concept [56], the dual hesitant fuzzy-based group decision making technique which assists the assessment of network information system security [57], a variety of distance measures for dual hesitant fuzzy sets [49], and the investigation of multiple criteria decision making problems based on Heronian mean, in which the attribute values are assumed in the form of interval-valued dual hesitant fuzzy information [59].

Definition 1.16 ([63]) Let X be a reference set, a dual hesitant fuzzy set (DHFS) $\mathbf{A}$ on X is defined in terms of two functions $u_{\mathbf{A}}(x)$ and $v_{\mathbf{A}}(x)$ as follows:

$$\mathbf{A} = \{\langle x, u_{\mathbf{A}}(x), v_{\mathbf{A}}(x)\rangle \mid x \in X\}, \tag{1.87}$$

where $u_{\mathbf{A}}(x)$ and $v_{\mathbf{A}}(x)$ are the sets of some different values in $[0, 1]$ and represent the possible membership degrees and non-membership degrees of the element $x \in X$ to $\mathbf{A}$, respectively.

Here, for all $x \in X$, if we suppose that $u_{\mathbf{A}}(x) = \bigcup_{\gamma_{\mathbf{A}} \in u_{\mathbf{A}}(x)}\{\gamma_{\mathbf{A}}\}$, $v_{\mathbf{A}}(x) = \bigcup_{\eta_{\mathbf{A}} \in v_{\mathbf{A}}(x)}\{\eta_{\mathbf{A}}\}$, ${\gamma_{\mathbf{A}}}^{+} \in u_{\mathbf{A}}^{+} = \bigcup_{x \in X} \max_{\gamma_{\mathbf{A}} \in u_{\mathbf{A}}(x)}\{\gamma_{\mathbf{A}}\}$ and ${\eta_{\mathbf{A}}}^{+} \in v_{\mathbf{A}}^{+} = \bigcup_{x \in X} \max_{\eta_{\mathbf{A}} \in v_{\mathbf{A}}(x)}\{\eta_{\mathbf{A}}\}$, then it would be concluded that

$$0 \leq \gamma_{\mathbf{A}},\ \eta_{\mathbf{A}} \leq 1, \quad 0 \leq {\gamma_{\mathbf{A}}}^{+} + {\eta_{\mathbf{A}}}^{+} \leq 1.$$

For the sake of simplicity, Zhu et al. [63] called the pair $h_{\mathbf{A}}(x) = (u_{\mathbf{A}}(x), v_{\mathbf{A}}(x))$ as the dual hesitant fuzzy element (DHFE).

Remark 1.7 Throughout this book, the set of all DHFSs on the reference set X is denoted by $\mathbb{DHFS}(X)$.

Remark 1.8 It should be mentioned that the concept of DHFS is also called in the literature [50] as the intuitionistic hesitant fuzzy set.

Before giving the definition of arithmetic operations for DHFSs, let us discuss more or less about the complement operator for DHFSs. The complement of a DHFS $\mathbf{A}$, denoted by $\mathbf{A}^c$ is defined in form of (see [63])

$$\begin{aligned}
\mathbf{A}^c &= \{\langle x, u_{\mathbf{A}}^c(x), v_{\mathbf{A}}^c(x)\rangle \mid x \in X\} \\
&= \begin{cases} \{\langle x, \bigcup_{\eta_{\mathbf{A}} \in v_{\mathbf{A}}(x)}\{\eta_{\mathbf{A}}\}, \bigcup_{\gamma_{\mathbf{A}} \in u_{\mathbf{A}}(x)}\{\gamma_{\mathbf{A}}\}\rangle | x \in X\}, & \text{if} \quad u_{\mathbf{A}} \neq \emptyset, v_{\mathbf{A}} \neq \emptyset; \\ \{\langle x, \bigcup_{\gamma_{\mathbf{A}} \in u_{\mathbf{A}}(x)}\{1 - \gamma_{\mathbf{A}}\}, \{\emptyset\}\rangle | x \in X\}, & \text{if} \quad u_{\mathbf{A}} \neq \emptyset, v_{\mathbf{A}} = \emptyset; \\ \{\langle x, \{\emptyset\}, \bigcup_{\eta_{\mathbf{A}} \in v_{\mathbf{A}}(x)}\{1 - \eta_{\mathbf{A}}\}\rangle | x \in X\}, & \text{if} \quad u_{\mathbf{A}} = \emptyset, v_{\mathbf{A}} \neq \emptyset. \end{cases}
\end{aligned}$$

Example 1.8 Let $X = \{x_1, x_2\}$ be the reference set, $h_{\mathbf{A}}(x_1) = (u_{\mathbf{A}}(x_1), v_{\mathbf{A}}(x_1)) = (\{0.2, 0.5\}, \{0.3\})$ and $h_{\mathbf{A}}(x_2) = (u_{\mathbf{A}}(x_2), v_{\mathbf{A}}(x_2)) = (\{0.3, 0.4\}, \{0.1, 0.6\})$ be the DHFEs of x_i $(i = 1, 2)$ in the set $\mathbf{A}$, respectively. Then $\mathbf{A}$ can be considered as a DHFS, i.e.,

$$\mathbf{A} = \{\langle x_1, \{0.2, 0.5\}, \{0.3\}\rangle,\ \langle x_2, \{0.3, 0.4\}, \{0.1, 0.6\}\rangle\}.$$

Note that for a given DHFE $h_{\mathbf{A}} \neq \emptyset$, if $u_{\mathbf{A}}$ and $v_{\mathbf{A}}$ possess only one value $\gamma_{\mathbf{A}}$ and $\eta_{\mathbf{A}}$, respectively, such that $0 \leq \gamma_{\mathbf{A}} + \eta_{\mathbf{A}} \leq 1$, then the DHFS reduces to an intuitionistic fuzzy set [54]. If $u_{\mathbf{A}}$ and $v_{\mathbf{A}}$ possess only one value $\gamma_{\mathbf{A}}$ and $\eta_{\mathbf{A}}$, respectively, such that $\gamma_{\mathbf{A}} + \eta_{\mathbf{A}} = 1$, or $u_{\mathbf{A}}$ has one value and $v_{\mathbf{A}} = \emptyset$, then the DHFS reduces to a fuzzy set. If $u_{\mathbf{A}} \neq \emptyset$ and $v_{\mathbf{A}} = \emptyset$, then the DHFS reduces to a HFS.

In what follows, we will borrow definition of some set and algebraic operations on DHFSs.

Definition 1.17 ([15]) Let X be a reference set, $\mathbf{A}_1$ and $\mathbf{A}_2$ be two DHFSs. We define

$$\begin{aligned}
\mathbf{A}_1 \cup \mathbf{A}_2 &= \bigcup_{h_{\mathbf{A}_1} \in \mathbf{A}_1, h_{\mathbf{A}_2} \in \mathbf{A}_2} h_{\mathbf{A}_1} \cup h_{\mathbf{A}_2} \\
&= \{\langle x, u_{\mathbf{A}_1}(x) \cup u_{\mathbf{A}_2}(x), v_{\mathbf{A}_1}(x) \cap v_{\mathbf{A}_2}(x)\rangle | x \in X\}, \qquad (1.88)
\end{aligned}$$

$$\mathbf{A}_1 \cap \mathbf{A}_2 = \bigcup_{h_{\mathbf{A}_1} \in \mathbf{A}_1, h_{\mathbf{A}_2} \in \mathbf{A}_2} h_{\mathbf{A}_1} \cap h_{\mathbf{A}_2}$$
$$= \{\langle x, u_{\mathbf{A}_1}(x) \cap u_{\mathbf{A}_2}(x), v_{\mathbf{A}_1}(x) \cup v_{\mathbf{A}_2}(x)\rangle | x \in X\}, \tag{1.89}$$

$$\mathbf{A}_1 \oplus \mathbf{A}_2 = \bigcup_{h_{\mathbf{A}_1} \in \mathbf{A}_1, h_{\mathbf{A}_2} \in \mathbf{A}_2} h_{\mathbf{A}_1} \oplus h_{\mathbf{A}_2}$$
$$= \{\langle x, u_{\mathbf{A}_1}(x) \oplus u_{\mathbf{A}_2}(x), v_{\mathbf{A}_1}(x) \otimes v_{\mathbf{A}_2}(x)\rangle | x \in X\}, \tag{1.90}$$

$$\mathbf{A}_1 \otimes \mathbf{A}_2 = \bigcup_{h_{\mathbf{A}_1} \in \mathbf{A}_1, h_{\mathbf{A}_2} \in \mathbf{A}_2} h_{\mathbf{A}_1} \otimes h_{\mathbf{A}_2}$$
$$= \{\langle x, u_{\mathbf{A}_1}(x) \otimes u_{\mathbf{A}_2}(x), v_{\mathbf{A}_1}(x) \oplus v_{\mathbf{A}_2}(x)\rangle | x \in X\}, \tag{1.91}$$

where the operations $u_{\mathbf{A}_1}(x) \circledast u_{\mathbf{A}_2}(x)$ and $v_{\mathbf{A}_1}(x) \circledast v_{\mathbf{A}_2}(x)$ coincide with those defined between HFSs in the preceding sections.

It is noteworthy that the above Zhu and Xu's definition of DHFE operations are much like that proposed by Farhadinia for DHFEs in [15].

Analogously, we can obtain the product and the sum operations of Algebraic, Einstein, Hamacher, and Frank for DHFEs as those proposed in Definition 1.7.

The division and subtraction operations of HFEs introduced in Definition 1.7 can be extended into that of DHFEs as follows:

Definition 1.18 ([15]) Given two DHFEs represented by $h_{\mathbf{A}_1} = (u_{\mathbf{A}_1}, v_{\mathbf{A}_1})$ and $h_{\mathbf{A}_2} = (u_{\mathbf{A}_2}, v_{\mathbf{A}_2})$, the division and the subtraction operations on the DHFEs, which are also DHFEs, can be described as follows:

$$h_{\mathbf{A}_1} \oslash h_{\mathbf{A}_2} = (u_{\mathbf{A}_1} \oslash u_{\mathbf{A}_2}, v_{\mathbf{A}_1} \ominus v_{\mathbf{A}_2}); \tag{1.92}$$
$$h_{\mathbf{A}_1} \ominus h_{\mathbf{A}_2} = (u_{\mathbf{A}_1} \ominus u_{\mathbf{A}_2}, v_{\mathbf{A}_1} \oslash v_{\mathbf{A}_2}), \tag{1.93}$$

where the right hand side operations can be taken as Algebraic, Einstein, Hamacher or Frank division and subtraction operations proposed in Definition 1.7.

1.7 Dual Hesitant Triangular Fuzzy Set

The concept of dual hesitant triangular fuzzy set, which is also known as hesitant triangular intuitionistic fuzzy [7], extends the concept of generalized trapezoidal hesitant fuzzy set introduced by Deli and Karaaslan [10]. Zhao et al. [61] gave the hesitant triangular fuzzy information together with a number of aggregating operators for aggregating all the related information of

(continued)

multiple attribute decision making problems. Chen and Huang [7] standardized hesitant triangular intuitionistic fuzzy aggregation operators, and they proposed several distance measures to explore their application in multiple criteria decision making.

Definition 1.19 ([7]) Let X be a reference set, a dual hesitant triangular fuzzy set (DHTFS) $\overline{A}$ on X is defined in terms of two functions $u_{\overline{A}}(x)$ and $v_{\overline{A}}(x)$ as follows:

$$\overline{A} = \{\langle x, u_{\overline{A}}(x), v_{\overline{A}}(x)\rangle \mid x \in X\}, \tag{1.94}$$

where $u_{\overline{A}}(x)$ and $v_{\overline{A}}(x)$ are the possible membership and non-membership degrees of the element $x \in X$ to $\overline{A}$, respectively. The dual hesitant triangular fuzzy element (DHTFE) $h_{\overline{A}}(x)$ can be also denoted by

$$\overline{A} = \left\{\left\langle x, \bigcup_{\langle(\gamma^L_{\overline{A}}, \gamma^M_{\overline{A}}, \gamma^U_{\overline{A}}), \gamma^u_{\overline{A}}, \gamma^v_{\overline{A}}\rangle \in h_{\overline{A}}(x)} \{\langle(\gamma^L_{\overline{A}}, \gamma^M_{\overline{A}}, \gamma^U_{\overline{A}}), \gamma^u_{\overline{A}}, \gamma^v_{\overline{A}}\rangle\}\right\rangle \mid x \in X\right\}, \tag{1.95}$$

where

$$u_{\overline{A}}(x) = \begin{cases} \frac{t-\gamma^L_{\overline{A}}}{\gamma^M_{\overline{A}}-\gamma^L_{\overline{A}}} \times \gamma^u_{\overline{A}} \text{ if } & \gamma^L_{\overline{A}} \le t \le \gamma^M_{\overline{A}}, \\ \frac{\gamma^U_{\overline{A}}-t}{\gamma^U_{\overline{A}}-\gamma^M_{\overline{A}}} \times \gamma^u_{\overline{A}} \text{ if } & \gamma^M_{\overline{A}} \le t \le \gamma^U_{\overline{A}}. \end{cases}$$

$$v_{\overline{A}}(x) = \begin{cases} \frac{\gamma^M_{\overline{A}}-t+(t-\gamma^L_{\overline{A}})\times\gamma^v_{\overline{A}}}{\gamma^M_{\overline{A}}-\gamma^L_{\overline{A}}} \text{ if } & \gamma^L_{\overline{A}} \le t \le \gamma^M_{\overline{A}}, \\ \frac{t-\gamma^M_{\overline{A}}+(\gamma^U_{\overline{A}}-t)\times\gamma^v_{\overline{A}}}{\gamma^U_{\overline{A}}-\gamma^M_{\overline{A}}} \text{ if } & \gamma^M_{\overline{A}} \le t \le \gamma^U_{\overline{A}}. \end{cases}$$

In the above representation, the values $\gamma^u_{\overline{A}}$ and $\gamma^v_{\overline{A}}$ stand, respectively, for the maximum degree of membership and the minimum degree of non-membership satisfying the conditions $\gamma^u_{\overline{A}}, \gamma^v_{\overline{A}} \in [0, 1]$ and $\gamma^u_{\overline{A}} + \gamma^v_{\overline{A}} \le 1$.

Remark 1.9 Throughout this book, the set of all DHTFSs on the reference set X is denoted by $\mathbb{DHTFS}(X)$.

Example 1.9 Let $X = \{x_1, x_2\}$ be the reference set, $h_{\overline{A}}(x_1) = \{\langle(0.2, 0, 3, 0.45), 0.4, 0.6\rangle, \langle(0.3, 0, 35, 0.45), 0.3, 0.5\rangle\}$ and $h_{\overline{A}}(x_2) = \{\langle(0.3, 0.5, 0.6), 0.7, 0.8\rangle\}$ be the DHTFEs of x_i $(i = 1, 2)$ to a set $\overline{A}$, respectively. Then $\overline{A}$ can be considered as an DHTFS, i.e.,

$$\overline{A} = \{\langle x_1, \langle(0.2, 0, 3, 0.45), 0.4, 0.6\rangle, \langle(0.3, 0, 35, 0.45), 0.3, 0.5\rangle, \\ \langle x_2, \langle(0.3, 0.5, 0.6), 0.7, 0.8\rangle\}.$$

Given three DHTFEs represented by $h_{\overline{A}} = \bigcup_{\langle(\gamma^L_{\overline{A}},\gamma^M_{\overline{A}},\gamma^U_{\overline{A}}),\gamma^u_{\overline{A}},\gamma^v_{\overline{A}}\rangle\in h_{\overline{A}}}\{\langle(\gamma^L_{\overline{A}}, \gamma^M_{\overline{A}}, \gamma^U_{\overline{A}}), \gamma^u_{\overline{A}}, \gamma^v_{\overline{A}}\rangle\}\rangle$, $h_{\overline{A}_1} = \bigcup_{\langle(\gamma^L_{\overline{A}_1},\gamma^M_{\overline{A}_1},\gamma^U_{\overline{A}_1}),\gamma^u_{\overline{A}_1},\gamma^v_{\overline{A}_1}\rangle\in h_{\overline{A}_1}}\{\langle(\gamma^L_{\overline{A}_1}, \gamma^M_{\overline{A}_1}, \gamma^U_{\overline{A}_1}), \gamma^u_{\overline{A}_1}, \gamma^v_{\overline{A}_1}\rangle\}\rangle$ and $h_{\overline{A}_2} = \bigcup_{\langle(\gamma^L_{\overline{A}_2},\gamma^M_{\overline{A}_2},\gamma^U_{\overline{A}_2}),\gamma^u_{\overline{A}_2},\gamma^v_{\overline{A}_2}\rangle\in h_{\overline{A}_2}}\{\langle(\gamma^L_{\overline{A}_2}, \gamma^M_{\overline{A}_2}, \gamma^U_{\overline{A}_2}), \gamma^u_{\overline{A}_2}, \gamma^v_{\overline{A}_2}\rangle\}\rangle$, some set and algebraic operations on the DHTFEs, which are also DHTFEs, can be described as follows (see, e.g., [7]):

$$\begin{aligned} &h_{\overline{A}_1} \oplus h_{\overline{A}_2} \\ &= \bigcup_{\substack{\langle(\gamma^L_{\overline{A}_1},\gamma^M_{\overline{A}_1},\gamma^U_{\overline{A}_1}),\gamma^u_{\overline{A}_1},\gamma^v_{\overline{A}_1}\rangle\in h_{\overline{A}_1} \\ \langle(\gamma^L_{\overline{A}_2},\gamma^M_{\overline{A}_2},\gamma^U_{\overline{A}_2}),\gamma^u_{\overline{A}_2},\gamma^v_{\overline{A}_2}\rangle\in h_{\overline{A}_2}}} \\ &\times\{\langle(\gamma^L_{\overline{A}_1} + \gamma^L_{\overline{A}_2}, \gamma^M_{\overline{A}_1} + \gamma^M_{\overline{A}_2}, \gamma^U_{\overline{A}_1} + \gamma^U_{\overline{A}_2}), \min\{\gamma^u_{\overline{A}_1}, \gamma^u_{\overline{A}_2}\}, \max\{\gamma^v_{\overline{A}_1}, \gamma^v_{\overline{A}_2}\}\rangle\}; \end{aligned} \tag{1.96}$$

$$\begin{aligned} &h_{\overline{A}_1} \otimes h_{\overline{A}_2} \\ &= \bigcup_{\substack{\langle(\gamma^L_{\overline{A}_1},\gamma^M_{\overline{A}_1},\gamma^U_{\overline{A}_1}),\gamma^u_{\overline{A}_1},\gamma^v_{\overline{A}_1}\rangle\in h_{\overline{A}_1} \\ \langle(\gamma^L_{\overline{A}_2},\gamma^M_{\overline{A}_2},\gamma^U_{\overline{A}_2}),\gamma^u_{\overline{A}_2},\gamma^v_{\overline{A}_2}\rangle\in h_{\overline{A}_2}}} \\ &\times\{\langle(\gamma^L_{\overline{A}_1}\gamma^L_{\overline{A}_2}, \gamma^M_{\overline{A}_1}\gamma^M_{\overline{A}_2}, \gamma^U_{\overline{A}_1}\gamma^U_{\overline{A}_2}), \min\{\gamma^u_{\overline{A}_1}, \gamma^u_{\overline{A}_2}\}, \max\{\gamma^v_{\overline{A}_1}, \gamma^v_{\overline{A}_2}\}\rangle\}; \end{aligned} \tag{1.97}$$

whenever both $h_{\overline{A}_1}, h_{\overline{A}_2}$ are positive, and moreover,

$$h^\lambda_{\overline{A}} = \bigcup_{\langle(\gamma^L_{\overline{A}},\gamma^M_{\overline{A}},\gamma^U_{\overline{A}}),\gamma^u_{\overline{A}},\gamma^v_{\overline{A}}\rangle\in h_{\overline{A}}} \{\langle(\gamma^{L\lambda}_{\overline{A}}, \gamma^{M\lambda}_{\overline{A}}, \gamma^{U\lambda}_{\overline{A}}), \gamma^u_{\overline{A}}, \gamma^v_{\overline{A}}\rangle\}; \quad \lambda > 0; \tag{1.98}$$

$$\lambda h_{\overline{A}} = \bigcup_{\langle(\gamma^L_{\overline{A}},\gamma^M_{\overline{A}},\gamma^U_{\overline{A}}),\gamma^u_{\overline{A}},\gamma^v_{\overline{A}}\rangle\in h_{\overline{A}}} \{\langle(\lambda\gamma^L_{\overline{A}}, \lambda\gamma^M_{\overline{A}}, \lambda\gamma^U_{\overline{A}}), \gamma^u_{\overline{A}}, \gamma^v_{\overline{A}}\rangle\}; \quad \lambda > 0. \tag{1.99}$$

Remark 1.10 In a recent work, Alcantud et al. [1] extended the concept of DHFS based on the combination of extended hesitant fuzzy set (EHFS) with DHFS. Indeed, the extended DHFS (EDHFS) of degree $\aleph$ is nothing else except the set of elements in the form of $\aleph$-tuple of DHFEs. In view of this vein, the set and algebraic operations on EDHFSs are the $\aleph$-tuple forms of those presented for DHFEs.

1.8 Interval-Valued Dual Hesitant Fuzzy Set

The concept of interval-valued hesitant fuzzy set was extended by Ju et al. [32] to the interval-valued dual hesitant fuzzy set using the assignment of same importance to the possible non-membership interval values in the interval-valued hesitant fuzzy set. In the sequel, Peng et al. [42] employed the concepts of Archimedean t-norm and Archimedean t-conorm to propose a number of interval-valued dual-hesitant fuzzy aggregation operators. Divers' type of interval-valued dual-hesitant fuzzy operators were developed by Zang et al. [59], Sarkar and Biswas [45] and Jiang et al. [31] for solving a variety of group decision making problems.

Definition 1.20 ([2]) Let X be a reference set. An interval-valued dual hesitant fuzzy set (IVDHFS) $\widetilde{\mathbf{A}}$ on X is defined in terms of two functions $u_{\widetilde{\mathbf{A}}}(x)$ and $v_{\widetilde{\mathbf{A}}}(x)$ as follows:

$$\widetilde{\mathbf{A}} = \{\langle x, u_{\widetilde{\mathbf{A}}}(x), v_{\widetilde{\mathbf{A}}}(x)\rangle \mid x \in X\}, \tag{1.100}$$

where

$$u_{\widetilde{\mathbf{A}}}(x) = \bigcup_{[\gamma^L_{\widetilde{\mathbf{A}}}, \gamma^U_{\widetilde{\mathbf{A}}}] \in u_{\widetilde{\mathbf{A}}}(x)} \{[\gamma^L_{\widetilde{\mathbf{A}}}, \gamma^U_{\widetilde{\mathbf{A}}}]\};$$

$$v_{\widetilde{\mathbf{A}}}(x) = \bigcup_{[\eta^L_{\widetilde{\mathbf{A}}}, \eta^U_{\widetilde{\mathbf{A}}}] \in v_{\widetilde{\mathbf{A}}}(x)} \{[\eta^L_{\widetilde{\mathbf{A}}}, \eta^U_{\widetilde{\mathbf{A}}}]\}$$

are some different interval values in [0, 1] and represent the possible interval membership degrees and interval non-membership degrees of the element $x \in X$, respectively.

Here, for all $x \in X$, if we suppose that $\gamma_{\widetilde{\mathbf{A}}}^{U+} = \bigcup_{x\in X} \max\{\gamma_{\widetilde{\mathbf{A}}}^{U}(x)\}$ and $\eta_{\widetilde{\mathbf{A}}}^{U+} = \bigcup_{x\in X} \max\{\eta_{\widetilde{\mathbf{A}}}^{U}(x)\}$, then it would be concluded that

$$0 \le \gamma_{\widetilde{\mathbf{A}}}^{L}, \gamma_{\widetilde{\mathbf{A}}}^{U}, \ \eta_{\widetilde{\mathbf{A}}}^{L}, \eta_{\widetilde{\mathbf{A}}}^{U} \le 1, \quad 0 \le \gamma_{\widetilde{\mathbf{A}}}^{U+} + \eta_{\widetilde{\mathbf{A}}}^{U+} \le 1.$$

For the sake of simplicity, $h_{\widetilde{\mathbf{A}}}(x) = (\bigcup_{[\gamma_{\widetilde{\mathbf{A}}}^{L},\gamma_{\widetilde{\mathbf{A}}}^{U}]\in u_{\widetilde{\mathbf{A}}}(x)}\{[\gamma_{\widetilde{\mathbf{A}}}^{L}, \gamma_{\widetilde{\mathbf{A}}}^{U}]\}, \bigcup_{[\eta_{\widetilde{\mathbf{A}}}^{L},\eta_{\widetilde{\mathbf{A}}}^{U}]\in v_{\widetilde{\mathbf{A}}}(x)}\{[\eta_{\widetilde{\mathbf{A}}}^{L}, \eta_{\widetilde{\mathbf{A}}}^{U}]\})$ is called the interval-valued dual hesitant fuzzy element (IVDHFE).

Remark 1.11 Throughout this book, the set of all IVDHFSs on the reference set X is denoted by $\mathbb{IVDHFS}(X)$.

Example 1.10 Let $X = \{x_1, x_2\}$ be the reference set, $h_{\widetilde{\mathbf{A}}}(x_1) = (\bigcup_{[\gamma_{\widetilde{\mathbf{A}}}^{L},\gamma_{\widetilde{\mathbf{A}}}^{U}]\in u_{\widetilde{\mathbf{A}}}(x_1)}\{[\gamma_{\widetilde{\mathbf{A}}}^{L}, \gamma_{\widetilde{\mathbf{A}}}^{U}]\}, \bigcup_{[\eta_{\widetilde{\mathbf{A}}}^{L},\eta_{\widetilde{\mathbf{A}}}^{U}]\in u_{\widetilde{\mathbf{A}}}(x_1)}\{[\eta_{\widetilde{\mathbf{A}}}^{L}, \eta_{\widetilde{\mathbf{A}}}^{U}]\}) = ([0.2, 0.5], [0.3, 0.3])$ and $h_{\widetilde{\mathbf{A}}}(x_2) = (\bigcup_{[\gamma_{\widetilde{\mathbf{A}}}^{L},\gamma_{\widetilde{\mathbf{A}}}^{U}]\in u_{\widetilde{\mathbf{A}}}(x_2)}\{[\gamma_{\widetilde{\mathbf{A}}}^{L}, \gamma_{\widetilde{\mathbf{A}}}^{U}]\}, \bigcup_{[\eta_{\widetilde{\mathbf{A}}}^{L},\eta_{\widetilde{\mathbf{A}}}^{U}]\in v_{\widetilde{\mathbf{A}}}(x_2)}\{[\eta_{\widetilde{\mathbf{A}}}^{L}, \eta_{\widetilde{\mathbf{A}}}^{U}]\}) = ([0.3, 0.4], [0.1, 0.6])$ be the IVDHFEs of x_i $(i = 1, 2)$ in the set $\widetilde{\mathbf{A}}$, respectively. Then $\widetilde{\mathbf{A}}$ can be considered as a IVDHFS, i.e.,

$$\widetilde{\mathbf{A}} = \{\langle x_1, [0.2, 0.5], [0.3, 0.3]\rangle, \ \langle x_2, [0.3, 0.4], [0.1, 0.6]\rangle\}.$$

For the IVDHFLEs
$h_{\widetilde{\mathbf{A}}} = (\bigcup_{[\gamma_{\widetilde{\mathbf{A}}}^{L},\gamma_{\widetilde{\mathbf{A}}}^{U}]\in u_{\widetilde{\mathbf{A}}}}\{[\gamma_{\widetilde{\mathbf{A}}}^{L}, \gamma_{\widetilde{\mathbf{A}}}^{U}]\}, \bigcup_{[\eta_{\widetilde{\mathbf{A}}}^{L},\eta_{\widetilde{\mathbf{A}}}^{U}]\in v_{\widetilde{\mathbf{A}}}}\{[\eta_{\widetilde{\mathbf{A}}}^{L}, \eta_{\widetilde{\mathbf{A}}}^{U}]\})$,
$h_{\widetilde{\mathbf{A}}_1} = (\bigcup_{[\gamma_{\widetilde{\mathbf{A}}_1}^{L},\gamma_{\widetilde{\mathbf{A}}_1}^{U}]\in u_{\widetilde{\mathbf{A}}_1}}\{[\gamma_{\widetilde{\mathbf{A}}_1}^{L}, \gamma_{\widetilde{\mathbf{A}}_1}^{U}]\}, \bigcup_{[\eta_{\widetilde{\mathbf{A}}_1}^{L},\eta_{\widetilde{\mathbf{A}}_1}^{U}]\in v_{\widetilde{\mathbf{A}}_1}}\{[\eta_{\widetilde{\mathbf{A}}_1}^{L}, \eta_{\widetilde{\mathbf{A}}_1}^{U}]\})$ and
$h_{\widetilde{\mathbf{A}}_2} = (\bigcup_{[\gamma_{\widetilde{\mathbf{A}}_2}^{L},\gamma_{\widetilde{\mathbf{A}}_2}^{U}]\in u_{\widetilde{\mathbf{A}}_2}}\{[\gamma_{\widetilde{\mathbf{A}}_2}^{L}, \gamma_{\widetilde{\mathbf{A}}_2}^{U}]\}, \bigcup_{[\eta_{\widetilde{\mathbf{A}}_2}^{L},\eta_{\widetilde{\mathbf{A}}_2}^{U}]\in v_{\widetilde{\mathbf{A}}_2}}\{[\eta_{\widetilde{\mathbf{A}}_2}^{L}, \eta_{\widetilde{\mathbf{A}}_2}^{U}]\})$ the following operations are defined (see [2]):

$$\begin{aligned} &h_{\widetilde{\mathbf{A}}_1} \oplus h_{\widetilde{\mathbf{A}}_2} \\ &= \left(\begin{array}{l} \bigcup\limits_{\substack{[\gamma_{\widetilde{\mathbf{A}}_1}^{L},\gamma_{\widetilde{\mathbf{A}}_1}^{U}]\in u_{\widetilde{\mathbf{A}}_1} \\ [\gamma_{\widetilde{\mathbf{A}}_2}^{L},\gamma_{\widetilde{\mathbf{A}}_2}^{U}]\in u_{\widetilde{\mathbf{A}}_2}}} \{[\gamma_{\widetilde{\mathbf{A}}_1}^{L} + \gamma_{\widetilde{\mathbf{A}}_2}^{L} - \gamma_{\widetilde{\mathbf{A}}_1}^{L} \times \gamma_{\widetilde{\mathbf{A}}_2}^{L}, \gamma_{\widetilde{\mathbf{A}}_1}^{U} + \gamma_{\widetilde{\mathbf{A}}_2}^{U} - \gamma_{\widetilde{\mathbf{A}}_1}^{U} \times \gamma_{\widetilde{\mathbf{A}}_2}^{U}]\}, \\ \bigcup\limits_{\substack{[\eta_{\widetilde{\mathbf{A}}_1}^{L},\eta_{\widetilde{\mathbf{A}}_1}^{U}]\in v_{\widetilde{\mathbf{A}}_1} \\ [\eta_{\widetilde{\mathbf{A}}_2}^{L},\eta_{\widetilde{\mathbf{A}}_2}^{U}]\in v_{\widetilde{\mathbf{A}}_2}}} \{[\eta_{\widetilde{\mathbf{A}}_1}^{L} \times \eta_{\widetilde{\mathbf{A}}_2}^{L}, \eta_{\widetilde{\mathbf{A}}_1}^{U} \times \eta_{\widetilde{\mathbf{A}}_2}^{U}]\} \end{array} \right); \end{aligned} \tag{1.101}$$

$$h_{\widetilde{\mathbf{A}}_1} \otimes h_{\widetilde{\mathbf{A}}_2}$$

$$= \left(\bigcup_{\substack{[\gamma^L_{\widetilde{\mathbf{A}}_1}, \gamma^U_{\widetilde{\mathbf{A}}_1}] \in u_{\widetilde{\mathbf{A}}_1} \\ [\gamma^L_{\widetilde{\mathbf{A}}_2}, \gamma^U_{\widetilde{\mathbf{A}}_2}] \in u_{\widetilde{\mathbf{A}}_2}}} \{[\gamma^L_{\widetilde{\mathbf{A}}_1} \times \gamma^L_{\widetilde{\mathbf{A}}_2}, \gamma^U_{\widetilde{\mathbf{A}}_1} \times \gamma^U_{\widetilde{\mathbf{A}}_2}]\}, \right.$$

$$\left. \bigcup_{\substack{[\eta^L_{\widetilde{\mathbf{A}}_1}, \eta^U_{\widetilde{\mathbf{A}}_1}] \in v_{\widetilde{\mathbf{A}}_1} \\ [\eta^L_{\widetilde{\mathbf{A}}_2}, \eta^U_{\widetilde{\mathbf{A}}_2}] \in v_{\widetilde{\mathbf{A}}_2}}} \{[\eta^L_{\widetilde{\mathbf{A}}_1} + \eta^L_{\widetilde{\mathbf{A}}_2} - \eta^L_{\widetilde{\mathbf{A}}_1} \times \eta^L_{\widetilde{\mathbf{A}}_2}, \eta^U_{\widetilde{\mathbf{A}}_1} + \eta^U_{\widetilde{\mathbf{A}}_2} - \eta^U_{\widetilde{\mathbf{A}}_1} \times \eta^U_{\widetilde{\mathbf{A}}_2}]\} \right); \tag{1.102}$$

$$\lambda h_{\widetilde{\mathbf{A}}} = \left(\bigcup_{[\gamma^L_{\widetilde{\mathbf{A}}}, \gamma^U_{\widetilde{\mathbf{A}}}] \in u_{\widetilde{\mathbf{A}}}} \{[1 - (1 - \gamma^L_{\widetilde{\mathbf{A}}})^\lambda, 1 - (1 - \gamma^U_{\widetilde{\mathbf{A}}})^\lambda]\}, \bigcup_{[\eta^L_{\widetilde{\mathbf{A}}}, \eta^U_{\widetilde{\mathbf{A}}}] \in v_{\widetilde{\mathbf{A}}}} \{[(\eta^L_{\widetilde{\mathbf{A}}})^\lambda, (\eta^U_{\widetilde{\mathbf{A}}})^\lambda]\} \right); \tag{1.103}$$

$$h_{\widetilde{\mathbf{A}}}^\lambda = \left(\bigcup_{[\gamma^L_{\widetilde{\mathbf{A}}}, \gamma^U_{\widetilde{\mathbf{A}}}] \in u_{\widetilde{\mathbf{A}}}} \{[(\gamma^L_{\widetilde{\mathbf{A}}})^\lambda, (\gamma^U_{\widetilde{\mathbf{A}}})^\lambda]\}, \bigcup_{[\eta^L_{\widetilde{\mathbf{A}}}, \eta^U_{\widetilde{\mathbf{A}}}] \in v_{\widetilde{\mathbf{A}}}} \{[1 - (1 - \eta^L_{\widetilde{\mathbf{A}}})^\lambda, 1 - (1 - \eta^U_{\widetilde{\mathbf{A}}})^\lambda]\} \right). \tag{1.104}$$

References

1. J.C.R. Alcantud, G. Santos-Garcia, X. Peng, J. Zhan, Dual extended hesitant fuzzy sets. Symmetry **11**, 5 (2019)
2. J. Ali, Z. Bashir, T. Rashid, Weighted interval-valued dual-hesitant fuzzy sets and its application in teaching quality assessment. Soft Comput. **25**, 3503–3530 (2020)
3. K. Atanassov, *Intuitionistic Fuzzy Sets, Theory and Applications* (Physica-Verlag, Heidelberg, 1999)
4. B. Bedregal, G. Beliakov, H. Bustince, T. Calvo, R. Mesiar, D. Paternain, A class of fuzzy multisets with a fixed number of memberships. Inf. Sci. **189**, 1–17 (2012)
5. B. Bedregal, R. Reiser, H. Bustince, C. Lopez-Molina, V. Torra, Aggregating functions for typical hesitant fuzzy elements and the action of automorphisms. Inf. Sci. **256**, 82–97 (2014)
6. G. Beliakov, A. Pradera, T. Calvo, *Aggregation Functions: A Guide for Practitioners* (Springer, Heidelberg, 2007)
7. J. Chen, X. Huang, Hesitant triangular intuitionistic fuzzy information and its application to multi-attribute decision making problem. J. Nonlinear Sci. Appl. **10**, 1012–1029 (2017)
8. N. Chen, Z. Xu, M. Xia, Interval-valued hesitant preference relations and their applications to group decision making. Knowl. Based Syst. **37**, 528–540 (2013)
9. I. Deli, A TOPSIS method by using generalized trapezoidal hesitant fuzzy numbers and application to a robot selection problem. IEEE Trans. Fuzzy Syst. **38**, 779–793 (2020)
10. I. Deli, F. Karaaslan, Generalized trapezoidal hesitant fuzzy numbers and their applications to multiple criteria decision making problems. Soft Comput. **25**, 1017–1032 (2021)

11. D. Dubois, H. Prade, *Fuzzy Sets and Systems: Theory and Applications* (Academic, New York, 1980)
12. B. Farhadinia, A novel method of ranking hesitant fuzzy values for multiple attribute decision-making problems. Int. J. Intell. Syst. **28**, 752–767 (2013)
13. B. Farhadinia, Information measures for hesitant fuzzy sets and interval-valued hesitant fuzzy sets. Inf. Sci. **240**, 129–144 (2013)
14. B. Farhadinia, Distance and similarity measures for higher order hesitant fuzzy sets. Knowl. Based Syst. **55**, 43–48 (2014)
15. B. Farhadinia, Correlation for dual hesitant fuzzy sets and dual interval-valued hesitant fuzzy sets. Int. J. Intell. Syst. **29**, 184–205 (2014)
16. B. Farhadinia, A series of score functions for hesitant fuzzy sets. Inf. Sci. **277**, 102–110 (2014)
17. B. Farhadinia, Study on division and subtraction operations for hesitant fuzzy sets, interval-valued hesitant fuzzy sets and typical dual hesitant fuzzy sets. J. Intell. Fuzzy Syst. **28**, 1393–1402 (2015)
18. B. Farhadinia, Multiple criteria decision-making methods with completely unknown weights in hesitant fuzzy linguistic term setting. Knowl. Based Syst. **93**, 135–144 (2016)
19. B. Farhadinia, Hesitant fuzzy set lexicographical ordering and its application to multi-attribute decision making. Inf. Sci. **327**, 233–245 (2016)
20. B. Farhadinia, Determination of entropy measures for the ordinal scale-based linguistic models. Inf. Sci. **369**, 63–79 (2016)
21. B. Farhadinia, A multiple criteria decision making model with entropy weight in an interval-transformed hesitant fuzzy environment. Cogn. Comput. **9**, 513–525 (2017)
22. B. Farhadinia, Improved correlation measures for hesitant fuzzy sets, in *2018 6th Iranian Joint Congress on Fuzzy and Intelligent Systems (CFIS)*. https://doi.org/10.1109/CFIS.2018.8336664
23. B. Farhadinia, E. Herrera-Viedma, Entropy measures for hesitant fuzzy linguistic term sets using the concept of interval-transformed hesitant fuzzy elements. Int. J. Fuzzy Syst. **20**, 2122–2134 (2018)
24. B. Farhadinia, E. Herrera-Viedma, Sorting of decision-making methods based on their outcomes using dominance-vector hesitant fuzzy-based distance. Soft Comput. **23**, 1109–1121 (2019)
25. B. Farhadinia, E. Herrera-Viedma, Multiple criteria group decision making method based on extended hesitant fuzzy sets with unknown weight information. Appl. Soft Comput. **78**, 310–323 (2019)
26. B. Farhadinia, Z.S Xu, Distance and aggregation-based methodologies for hesitant fuzzy decision making. Cogn. Comput. **9**, 81–94 (2017)
27. B. Farhadinia, Z.S. Xu, Novel hesitant fuzzy linguistic entropy and cross-entropy measures in multiple criteria decision making. Appl. Intell. **48**, 3915–3927 (2018)
28. B. Farhadinia, Z.S. Xu, Ordered weighted hesitant fuzzy information fusion-based approach to multiple attribute decision making with probabilistic linguistic term sets. Fund. Inform. **159**, 361–383 (2018)
29. B. Farhadinia, Z.S. Xu, Hesitant fuzzy information measures derived from T-norms and S-norms. Iran. J. Fuzzy Syst. **15**, 157–175 (2018)
30. B. Farhadinia, Z.S Xu, An extended hesitant group decision-making technique based on the prospect theory for emergency situations. Iran. J. Fuzzy Syst. **17**, 51–68 (2020)
31. C. Jiang, S. Jiang, J. Chen, Interval-valued dual hesitant fuzzy Hamacher aggregation operators for multiple attribute decision making. J. Syst. Sci. Inf. **7**, 227–256 (2019)
32. Y. Ju, X. Liu, S. Yang, Interval-valued dual hesitant fuzzy aggregation operators and their application to multiple attribute decision making. Int. J. Intell. Syst. **27**, 1203–1218 (2014)
33. G.J. Klir, B. Yuan, *Fuzzy Sets and Fuzzy Logic-Theory and Applications* (Prentice-Hall, Upper Saddle River, 1995)
34. L. Li, D. Peng, Interval-valued hesitant fuzzy Hamacher synergetic weighted aggregation operators and their application to shale gas areas selection. Math. Probl. Eng. **2014**, 1–15 (2014)

35. H.C. Liao, Z.S. Xu, A VIKOR-based method for hesitant fuzzy multi-criteria decision making. Fuzzy Optim. Decis. Making **12**, 373–392 (2013)
36. H.C. Liao, Z.S. Xu, Subtraction and division operations over hesitant fuzzy sets. J. Intell. Fuzzy Syst. **27**, 65–72 (2014)
37. J.M. Mendel, *Rule-Based Fuzzy Logic Systems: Introduction and New Directions* (Prentice-Hall, Upper Saddle River, 2001)
38. S. Miyamoto, Multisets and fuzzy multisets, in *Soft Computing and Human-Centered Machines*, ed. by Z.Q. Liu, S. Miyamoto (Springer, Berlin, 2000), pp. 9–33
39. X. Peng, Hesitant trapezoidal fuzzy aggregation operators based on Archimedean t-norm and t-conorm and their application in MADM with completely unknown weight information. Int. J. Uncertain. Quantif. **7**, 475–510 (2017)
40. D.H. Peng, Ch.Y. Gao, Zh.F. Gao, Generalized hesitant fuzzy synergetic weighted distance measures and their application to multiple criteria decision making. Appl. Math. Model. **37**, 5837–5850 (2013)
41. D. Peng, T. Wang, C. Gao, H. Wang, Continuous hesitant fuzzy aggregation operators and their application to decision making under interval-valued hesitant fuzzy setting. Sci. World J. (2014). https://doi.org/10.1155/2014/897304
42. X. Peng, J. Dai, L. Liu, Interval-valued dual hesitant fuzzy information aggregation and its application in multiple attribute decision making. Int. J. Uncertain. Quantif. **8**, 361–382 (2018)
43. T. Rashid, S.M. Husnine, Multicriteria group decision making by using trapezoidal valued hesitant fuzzy sets. Sci. World J. Article ID 304834. 487 (2014). https://doi.org/10.1155/2014-304834
44. R.M. Rodriguez, L. Martinez, F. Herrera, A multicriteria linguistic decision making model dealing with comparative terms, in *Eurofuse* (2011), pp. 229–241
45. A. Sarkar, A. Biswas, On developing interval-valued dual hesitant fuzzy Bonferroni mean aggregation operator and their application to multicriteria decision making, in *International Conference on Computational Intelligence, Communications, and Business Analytics CICBA 2018: Computational Intelligence, Communications, and Business Analytics* (2018), pp. 27–46
46. J. Shi, C. Meng, Y. Liu, Approach to multiple attribute decision making based on the intelligence computing with hesitant triangular fuzzy information and their application. J. Intell. Fuzzy Syst. **27**, 701–707 (2014)
47. V. Torra, Hesitant fuzzy sets. Int. J. Intell. Syst. **25**, 529–539 (2010)
48. I.B. Turksen, Interval valued fuzzy sets based on normal forms. Fuzzy Sets Syst. **20**, 191–210 (1986)
49. L. Wang, S. Xu, Q. Wang, M. Ni, Distance and similarity measures of dual hesitant fuzzy sets with their applications to multiple attribute decision making, in *2014 International Conference on Progress in Informatics and Computing (PIC)* (IEEE, Piscataway, 2014), pp. 88–92
50. G. Wei, Hesitant fuzzy prioritized operators and their application to multiple attribute decision making. Knowl. Based Syst. **31**, 176–182 (2012)
51. G. Wei, X. Zhao, R. Lin, Some hesitant interval-valued fuzzy aggregation operators and their applications to multiple attribute decision making. Knowl. Based Syst. **46**, 43–53 (2013)
52. M.M. Xia, Z.S. Xu, Hesitant fuzzy information aggregation in decision making. Int. J. Approx. Reason. **52**, 395–407 (2011)
53. Z.S. Xu, *Linguistic Decision Making: Theory and Methods* (Science Press, Beijing, 2012)
54. Z.S. Xu, M.M. Xia, Distance and similarity measures for hesitant fuzzy sets. Inf. Sci. **181**, 2128–2138 (2011)
55. D. Yu, Triangular hesitant fuzzy set and its application to teaching quality evaluation. J. Inf. Comput. Sci. **10**, 1925–1934 (2013)
56. D. Yu, D.F. Li, J.M. Merigo, Dual hesitant fuzzy group decision making method and its application to supplier selection. Int. J. Mach. Learn. Cybern. **7**, 819–831 (2016)
57. D. Yu, J.M. Merigo, Y. Xu, Group decision making in information systems security assessment using dual hesitant fuzzy set. Int. J. Intell. Syst. **31**, 786–812 (2016)
58. L.A. Zadeh, Fuzzy sets. Inf. Comput. **8**, 338–353 (1965)

59. Y. Zang, X. Zhao, S. Li, Interval-valued dual hesitant fuzzy Heronian mean aggregation operators and their application to multi-attribute decision making. Int. J. Comput. Intell. Appl. **17**, 1850005 (2018)
60. X. Zhang, Z. Xu, M. Liu, Hesitant trapezoidal fuzzy QUALIFLEX method and its application in the evaluation of green supply chain initiatives. Sustainability **8** (2016). https://doi.org/10.3390/su8090952
61. X.F. Zhao, R. Lin, G. Wei, Hesitant triangular fuzzy information aggregation based on Einstein operations and their application to multiple attribute decision making. Expert Syst. Appl. **41**, 1086–1094 (2014)
62. B. Zhu, Z.S. Xu, Extended hesitant fuzzy sets. Technol. Econ. Dev. Econ. **22**, 100–121 (2016)
63. B. Zhu, Z.S. Xu, M.M. Xia, Dual hesitant fuzzy sets. J. Appl. Math. (2012). https://doi.org/10.1155/2012/879629
64. B. Zhu, Z.S. Xu, M.M. Xia, Hesitant fuzzy geometric Bonferroni means. Inf. Sci. **205**, 72–85 (2012)

Chapter 2
Hesitant Fuzzy Linguistic Term Set

Abstract In this chapter, we first deal with the concept of hesitant fuzzy linguistic term set that reflects the inconsistency, hesitancy, and uncertainty of experts. An extension of hesitant fuzzy linguistic term set, which is known as extended hesitant fuzzy linguistic term set, is represented and a number of operations are reviewed for such a concept. Then, by introducing the interval-valued hesitant fuzzy linguistic term set concept in the third section, we reveal another form of hesitant fuzzy linguistic term set extension. To address the proportional form of hesitant fuzzy linguistic term sets, we create the skeleton of next section. Hesitant fuzzy uncertain linguistic set is the topic being considered in the fifth section. One of the main extensions of hesitant fuzzy linguistic term set is the dual form by which we take much more information into account given by decision makers. Two other generalized forms of hesitant fuzzy linguistic term set including dual hesitant fuzzy linguistic triangular set and interval-valued dual hesitant fuzzy linguistic set are the issues lie at the next arguments in this chapter.

2.1 Hesitant Fuzzy Linguistic Term Set

Rodriguez et al. [14, 15] represented the concept of hesitant fuzzy linguistic term set which permits a linguistic variable to own various linguistic terms. Under hesitant fuzzy linguistic situation, Liao et al. [8] developed a satisfactory based decision making technique. Wei et al. [25] extended the comparison rule for hesitant fuzzy linguistic term sets and studied the theory of hesitant fuzzy linguistic term set aggregation. By the help of pessimistic and optimistic attitudes of decision makers, Chen and Hong [2] represented a multiple criteria linguistic decision technique. Zhu and Xu [33] introduced the concept of hesitant fuzzy linguistic preference relation, and moreover, they investigated its consistency. By taking comparative linguistic expressions into account, Liu et al. [11] improved the additive consistency of the hesitant fuzzy linguistic preference relations.

B. Farhadinia, *Hesitant Fuzzy Set*, Computational Intelligence Methods
and Applications, https://doi.org/10.1007/978-981-16-7301-6_2

In decision making problems with linguistic information, experts usually feel more comfortable to express their opinions by linguistic variables (or linguistic terms) because this approach is more realistic and it is close to the human cognitive processes. If we take into consideration the following finite and totally ordered discrete linguistic term set

$$\mathfrak{S} = \{s_\alpha \mid \alpha = -\tau, \ldots, -1, 0, 1, \ldots, \tau\}, \tag{2.1}$$

where τ is a positive integer, then we find that s_α represents a possible value for a linguistic variable.
Notice that the linguistic term set is sometimes considered as

$$\mathfrak{S} = \{s_\alpha \mid \alpha = 0, 1, \ldots, \tau\}, \tag{2.2}$$

which is not necessarily symmetrical about the central of linguistic term sets.
As a notational convention, we assume that the mid-linguistic label s_0 represents an assessment of *indifference*, and the remaining linguistic labels are symmetrically located around s_0. It is necessary that the totally ordered linguistic term set $\mathfrak{S}$ satisfies the following characteristics:

1. $s_\alpha < s_\beta$ if and only if $\alpha < \beta$;
2. The negation operator is defined as $N(s_\alpha) = s_{-\alpha}$.

For any two linguistic terms $s_\alpha, s_\beta \in \mathfrak{S}$, the following properties hold true [29]:

$$s_\alpha \oplus s_\beta = s_{\alpha+\beta}; \tag{2.3}$$

$$s_\alpha \oplus s_\beta = s_\beta \oplus s_\alpha; \tag{2.4}$$

$$\lambda s_\alpha = s_{\lambda\alpha}; \tag{2.5}$$

$$(\lambda_1 + \lambda_2)s_\alpha = \lambda_1 s_\alpha \oplus \lambda_2 s_\alpha; \tag{2.6}$$

$$\lambda(s_\alpha \oplus s_\beta) = \lambda s_\alpha \oplus \lambda s_\beta, \tag{2.7}$$

where $0 \le \lambda, \lambda_1, \lambda_2 \le 1$.

By the inspiration of the idea of HFS [17], Rodriguez et al. [14] introduced the hesitant fuzzy linguistic term set (HFLTS) to overcome some difficulties observed in a qualitative circumstance where a decision maker may hesitate between several terms at the same time, or he/she needs a complex linguistic term instead of a single linguistic term to assess a linguistic variable. Continuing that work, Liao et al. [7] refined the concept of HFLTS mathematically as follows:

Definition 2.1 ([7]) Let X be a reference set, and $\mathfrak{S} = \{s_\alpha \mid \alpha = -\tau, \ldots, -1, 0, 1, \ldots, \tau\}$ be a linguistic term set. A hesitant fuzzy linguistic term set (HFLTS) on X is mathematically shown in terms of

$$A^{\mathfrak{S}} = \{\langle x, h_{A^{\mathfrak{S}}}(x)\rangle \mid x \in X\}. \tag{2.8}$$

Here, $h_{A^{\mathfrak{S}}}(x)$ is a set of some possible values in the linguistic term set $\mathfrak{S}$ and can be characterized by

$$h_{A^{\mathfrak{S}}}(x) = \bigcup_{s_\alpha(x) \in \mathfrak{S}(x)} \{s_\alpha(x)\}. \tag{2.9}$$

Moreover, $h_{A^{\mathfrak{S}}}(x)$ is called the hesitant fuzzy linguistic element (HFLE) and we denote $h_{A^{\mathfrak{S}}}(x)$ briefly by $h_{A^{\mathfrak{S}}}$.

Example 2.1 Suppose that an expert is invited to evaluate the approximate speed of three cars x_1, x_2, and x_3. Note that this criterion is qualitative, and therefore it should be described by linguistic terms instead of crisp values. The expert's judgements over these three cars are given in the form of linguistic expressions as: $\{\langle x_1, \{fast,\ very\ fast\}\rangle, \langle x_2, \{very\ slow,\ slow,\ slightly\ slow,\ average\}\rangle, \langle x_3, \{very\ fast\}\rangle\}$, which establishes a HFLTS as:

$$A^{\mathfrak{S}} = \{\langle x_1, h_{A^{\mathfrak{S}}}(x_1) = \{s_2,\ s_3\}\rangle,$$
$$\langle x_2, h_{A^{\mathfrak{S}}}(x_2) = \{s_{-3},\ s_{-2},\ s_{-1},\ s_0\}\rangle, \langle x_3, h_{A^{\mathfrak{S}}}(x_3) = \{s_3\}\rangle\}.$$

Let $\mathfrak{S} = \{s_\alpha \mid \alpha = 0, 1, \ldots, \tau\}$ be the linguistic term set, and $h_{A^{\mathfrak{S}}} = \bigcup_{s_\alpha \in \mathfrak{S}} \{s_\alpha\}$, $h_{A_1{}^{\mathfrak{S}}} = \bigcup_{s_{\alpha_1} \in \mathfrak{S}} \{s_{\alpha_1}\}$ and $h_{A_2{}^{\mathfrak{S}}} = \bigcup_{s_{\alpha_2} \in \mathfrak{S}} \{s_{\alpha_2}\}$ be three HFLEs. Some set operations on HFLEs can be described as follows (see, e.g., [14]):

$$h^c_{A^{\mathfrak{S}}} = \bigcup_{s_\alpha \in \mathfrak{S}} \{s_{\tau-\alpha}\}; \tag{2.10}$$

$$h_{A_1{}^{\mathfrak{S}}} \cup h_{A_2{}^{\mathfrak{S}}} = \bigcup_{s_{\alpha_1}, s_{\alpha_2} \in \mathfrak{S}} \{s_{\max\{\alpha_1,\alpha_2\}}\}; \tag{2.11}$$

$$h_{A_1{}^{\mathfrak{S}}} \cap h_{A_2{}^{\mathfrak{S}}} = \bigcup_{s_{\alpha_1}, s_{\alpha_2} \in \mathfrak{S}} \{s_{\min\{\alpha_1,\alpha_2\}}\}. \tag{2.12}$$

Remark 2.1 By considering this fact that a group of experts may think of several possible linguistic values or richer expressions than a single term for evaluating an alternative in group decision making, Wang [18] discussed linguistic terms involved in an expression which is derived by the multiple decision makers being not always consecutive. Therefore, Wang generalized the concept of HFLTS to extended hesitant fuzzy linguistic term set (EHFLTS) in which the family of linguistic term sets is to be taken an ordered subset.

2.2 Interval-Valued Hesitant Fuzzy Linguistic Term Set

Wang et al. [20] proposed the concept of interval-valued hesitant fuzzy linguistic term set to combine the advantages of both linguistic term sets and interval-valued hesitant fuzzy sets. Indeed, the main advantage of interval-valued hesitant fuzzy linguistic term set is that it can describe two fuzzy criteria of an object, namely a linguistic term and an interval-valued hesitant fuzzy element. The former provides an evaluation value, such as "good" or "excellent", and the latter describes the hesitancy for the given evaluation value and denotes the interval-valued membership degrees associated with the specific linguistic term. Meng et al. [12] in a re-phrased concept, which was called linguistic interval hesitant fuzzy set, implemented interval-valued hesitant fuzzy linguistic term set concept to cope with the situations where the membership degrees of linguistic terms are intervals rather than real numbers.

Definition 2.2 ([20]) Let X be a reference set, and $\mathfrak{S} = \{s_\alpha \mid \alpha = -\tau, \ldots, -1, 0, 1, \ldots, \tau\}$ be a linguistic term set. An interval-valued hesitant fuzzy linguistic term set (IVHFLTS) on X is mathematically shown in terms of

$$\widetilde{A}^{\mathfrak{S}} = \{\langle x, h_{\widetilde{A}^{\mathfrak{S}}}(x)\rangle \mid x \in X\}. \tag{2.13}$$

Here, $h_{\widetilde{A}^{\mathfrak{S}}}(x)$ is a set of some possible pairs of linguistic term set $\mathfrak{S}$ together with interval values which can be characterized by

$$h_{\widetilde{A}^{\mathfrak{S}}}(x) = \langle s_\alpha(x), \bigcup_{[\gamma^L_{\widetilde{A}^{\mathfrak{S}}}, \gamma^U_{\widetilde{A}^{\mathfrak{S}}}] \in h_{\widetilde{A}^{\mathfrak{S}}}(x)} \{[\gamma^L_{\widetilde{A}^{\mathfrak{S}}}, \gamma^U_{\widetilde{A}^{\mathfrak{S}}}]\}\rangle, \tag{2.14}$$

where $s_\alpha(x) \in \mathfrak{S}$ denotes the possible value for a linguistic variable involved in $h_{\widetilde{A}^{\mathfrak{S}}}(x)$.

For any $x \in X$, we call $h_{\widetilde{A}^{\mathfrak{S}}}(x)$ an interval-valued hesitant fuzzy linguistic term element (IVHFLTE).

Remark 2.2 Throughout this book, the set of all IVHFLTSs on the reference set X is denoted by $\mathbb{IVHFLTS}(X)$.

Example 2.2 Let $X = \{x_1, x_2\}$ be the reference set, $h_{\widetilde{A}^{\mathfrak{S}}}(x_1) = \langle s_2, \{[0.2, 0, 3], [0.4, 0.6], [0.5, 0.6]\}\rangle$ and $h_{\widetilde{A}^{\mathfrak{S}}}(x_2) = \langle s_1, \{[0.3, 0.5], [0.4, 0.7]\}\rangle$ be the IVHFLTEs of x_i $(i = 1, 2)$ to a set $\widetilde{A}^{\mathfrak{S}}$, respectively. Then $\widetilde{A}^{\mathfrak{S}}$

can be considered as an IVHFLTS, i.e.,

$$\begin{aligned}\widetilde{A}^{\mathfrak{S}} &= \{\langle x_1, \langle s_2, \{[0.2, 0, 3], [0.4, 0.6], [0.5, 0.6]\}\rangle\rangle, \\ &\quad \langle x_2, \langle s_1, \{[0.3, 0.5], [0.4, 0.7]\}\rangle\rangle\}.\end{aligned}$$

Given three IVHFLTEs represented by $h_{\widetilde{A}^{\mathfrak{S}}}$, $h_{\widetilde{A}_1^{\mathfrak{S}}}$ and $h_{\widetilde{A}_2^{\mathfrak{S}}}$, some set and algebraic operations on the IVHFLTEs, which are also IVHFLTEs, can be described as follows (see, e.g., [20]):

$$h^c_{\widetilde{A}^{\mathfrak{S}}} = \left\langle s_{-\alpha}, \bigcup_{[\gamma^L_{\widetilde{A}^{\mathfrak{S}}}, \gamma^U_{\widetilde{A}^{\mathfrak{S}}}] \in h_{\widetilde{A}^{\mathfrak{S}}}} \{[1 - \gamma^U_{\widetilde{A}^{\mathfrak{S}}}, 1 - \gamma^L_{\widetilde{A}^{\mathfrak{S}}}]\} \right\rangle, \tag{2.15}$$

$$\begin{aligned} & h_{\widetilde{A}_1^{\mathfrak{S}}} \cup h_{\widetilde{A}_2^{\mathfrak{S}}} \\ &= \left\langle \max\{s_{\alpha_1}, s_{\alpha_2}\}, \bigcup_{\substack{[\gamma^L_{\widetilde{A}_1^{\mathfrak{S}}}, \gamma^U_{\widetilde{A}_1^{\mathfrak{S}}}] \in h_{\widetilde{A}_1^{\mathfrak{S}}} \\ [\gamma^L_{\widetilde{A}_2^{\mathfrak{S}}}, \gamma^U_{\widetilde{A}_2^{\mathfrak{S}}}] \in h_{\widetilde{A}_2^{\mathfrak{S}}}}} \left\{ \left[\max\{\gamma^L_{\widetilde{A}_1^{\mathfrak{S}}}, \gamma^L_{\widetilde{A}_2^{\mathfrak{S}}}\}, \right.\right.\right. \\ & \quad \left.\left.\left. \max\{\gamma^U_{\widetilde{A}_1^{\mathfrak{S}}}, \gamma^U_{\widetilde{A}_2^{\mathfrak{S}}}\} \right] \right\} \right\rangle; \end{aligned} \tag{2.16}$$

$$\begin{aligned} & h_{\widetilde{A}_1^{\mathfrak{S}}} \cup h_{\widetilde{A}_2^{\mathfrak{S}}} \\ &= \left\langle \max\{s_{\alpha_1}, s_{\alpha_2}\}, \bigcup_{\substack{[\gamma^L_{\widetilde{A}_1^{\mathfrak{S}}}, \gamma^U_{\widetilde{A}_1^{\mathfrak{S}}}] \in h_{\widetilde{A}_1^{\mathfrak{S}}} \\ [\gamma^L_{\widetilde{A}_2^{\mathfrak{S}}}, \gamma^U_{\widetilde{A}_2^{\mathfrak{S}}}] \in h_{\widetilde{A}_2^{\mathfrak{S}}}}} \left\{ \left[\max\{\gamma^L_{\widetilde{A}_1^{\mathfrak{S}}}, \gamma^L_{\widetilde{A}_2^{\mathfrak{S}}}\}, \right.\right.\right. \\ & \quad \left.\left.\left. \max\{\gamma^U_{\widetilde{A}_1^{\mathfrak{S}}}, \gamma^U_{\widetilde{A}_2^{\mathfrak{S}}}\} \right] \right\} \right\rangle; \end{aligned} \tag{2.17}$$

$$\begin{aligned} & h_{\widetilde{A}_1^{\mathfrak{S}}} \oplus h_{\widetilde{A}_2^{\mathfrak{S}}} \\ &= \left\langle s_{\alpha_1+\alpha_2}, \bigcup_{\substack{[\gamma^L_{\widetilde{A}_1^{\mathfrak{S}}}, \gamma^U_{\widetilde{A}_1^{\mathfrak{S}}}] \in h_{\widetilde{A}_1^{\mathfrak{S}}} \\ [\gamma^L_{\widetilde{A}_2^{\mathfrak{S}}}, \gamma^U_{\widetilde{A}_2^{\mathfrak{S}}}] \in h_{\widetilde{A}_2^{\mathfrak{S}}}}} \left\{ \left[\gamma^L_{\widetilde{A}_1^{\mathfrak{S}}} + \gamma^L_{\widetilde{A}_2^{\mathfrak{S}}} - \gamma^L_{\widetilde{A}_1^{\mathfrak{S}}} \times \gamma^L_{\widetilde{A}_2^{\mathfrak{S}}}, \gamma^U_{\widetilde{A}_1^{\mathfrak{S}}} \right.\right.\right. \\ & \quad \left.\left.\left. + \gamma^U_{\widetilde{A}_2^{\mathfrak{S}}} - \gamma^U_{\widetilde{A}_1^{\mathfrak{S}}} \times \gamma^U_{\widetilde{A}_2^{\mathfrak{S}}} \right] \right\} \right\rangle; \end{aligned} \tag{2.18}$$

$$h_{\tilde{A}_1^{\mathfrak{S}}} \otimes h_{\tilde{A}_2^{\mathfrak{S}}} = \left\langle s_{\alpha_1 \times \alpha_2}, \bigcup_{\substack{[\gamma^L_{\tilde{A}_1^{\mathfrak{S}}}, \gamma^U_{\tilde{A}_1^{\mathfrak{S}}}] \in h_{\tilde{A}_1^{\mathfrak{S}}} \\ [\gamma^L_{\tilde{A}_2^{\mathfrak{S}}}, \gamma^U_{\tilde{A}_2^{\mathfrak{S}}}] \in h_{\tilde{A}_2^{\mathfrak{S}}}}} \{[\gamma^L_{\tilde{A}_1^{\mathfrak{S}}} \times \gamma^L_{\tilde{A}_2^{\mathfrak{S}}}, \gamma^U_{\tilde{A}_1^{\mathfrak{S}}} \times \gamma^U_{\tilde{A}_2^{\mathfrak{S}}}]\} \right\rangle; \tag{2.19}$$

$$h_{\tilde{A}^{\mathfrak{S}}}{}^{\lambda} = \left\langle s_{\alpha^{\lambda}}, \bigcup_{[\gamma^L_{\tilde{A}^{\mathfrak{S}}}, \gamma^U_{\tilde{A}^{\mathfrak{S}}}] \in h_{\tilde{A}^{\mathfrak{S}}}} \{[(\gamma^L_{\tilde{A}^{\mathfrak{S}}})^{\lambda}, (\gamma^U_{\tilde{A}^{\mathfrak{S}}})^{\lambda}]\} \right\rangle, \quad \lambda > 0; \tag{2.20}$$

$$\lambda h_{\tilde{A}^{\mathfrak{S}}} = \left\langle s_{\lambda\alpha}, \bigcup_{[\gamma^L_{\tilde{A}^{\mathfrak{S}}}, \gamma^U_{\tilde{A}^{\mathfrak{S}}}] \in h_{\tilde{A}^{\mathfrak{S}}}} \{[1-(1-\gamma^L_{\tilde{A}^{\mathfrak{S}}})^{\lambda}, 1-(1-\gamma^U_{\tilde{A}^{\mathfrak{S}}})^{\lambda}]\} \right\rangle, \ \lambda > 0. \tag{2.21}$$

2.3 Proportional Hesitant Fuzzy Linguistic Term Set

Wang and Hao [21] introduced the proportional two-tuple model that enables experts to assign symbolic proportions to two successive linguistic terms. Then, Zhang et al. [31] included the proportional information into the model of linguistic representation. Motivated by the idea of Zhang et al. [31], Wu and Xu [27] represented a possibility distribution-based technique for addressing multiple criteria group decision making with hesitant fuzzy linguistic information. In order to propose a linguistic representation model which considers simultaneously the hesitant linguistic assessment of experts and the proportional information of each linguistic term under a group decision making environment, Chen et al. [1] introduced the notion of proportional hesitant fuzzy linguistic fuzzy term set.

Definition 2.3 ([1]) Let $\mathfrak{S} = \{s_\alpha \mid \alpha = 0, 1, \ldots, \tau\}$ be a finite and totally ordered linguistic term set with odd cardinality. A proportional hesitant fuzzy linguistic term set (PHFLTS) on X is mathematically shown in terms of

$$h_{\underline{A}^{\mathfrak{S}}} = \bigcup_{s_\alpha \in \mathfrak{S}, p_\alpha \in P_{\underline{A}^{\mathfrak{S}}}} \{\langle s_\alpha, p_\alpha \rangle\}. \tag{2.22}$$

Here, $P_{\underset{\sim}{A}^{\mathfrak{S}}} = (p_0, p_1, \ldots, p_\alpha, \ldots, p_\tau)$ denotes a vector which represents the proportions for linguistic terms s_α's. The value p_α implies the degree of possibility that the alternative carries an assessment value s_α provided by a group of experts. It needs to be considered that $0 \leq p_\alpha \leq 1$ satisfies $\sum_{\alpha=0}^{\tau} p_\alpha = 1$.

By considering the concepts HFLTS [14], which rolls as an ordered finite subset of consecutive linguistic terms in $\mathfrak{S}$, and extended HFLTS (EHFLTS) [18], which rolls as an ordered subset of the linguistic terms of $\mathfrak{S}$, we conclude that the HFLTS elements are those provided by an expert for indicating his/her hesitancy with several possible values for a linguistic variable. This is while, the EHFLTS elements are those provided by a group of experts whose hesitancy is to assess the possible values for a linguistic variable. By contrast, the PHFLTS elements can be provided by either a group of experts who are hesitant to assess the possible values for a linguistic variable or a group of experts who assess a linguistic variable with a single linguistic term.

Remark 2.3 Throughout this book, the set of all PHFLTSs is denoted by $\mathbb{PHFLTS}(X)$.

Given the linguistic term set $\mathfrak{S} = \{s_\alpha \mid \alpha = 0, 1, \ldots, \tau\}$ and three PHFLTSs be represented by $h_{\underset{\sim}{A}^{\mathfrak{S}}} = \bigcup_{s_\alpha \in \mathfrak{S}, p_\alpha \in P_{\underset{\sim}{A}^{\mathfrak{S}}}} \{\langle s_\alpha, p_\alpha\rangle\}$, $h_{\underset{\sim}{A}_1^{\mathfrak{S}}} = \bigcup_{s_\alpha \in \mathfrak{S}, p_\alpha^1 \in P_{\underset{\sim}{A}_1^{\mathfrak{S}}}} \{\langle s_\alpha, p_\alpha^1\rangle\}$ and $h_{\underset{\sim}{A}_2^{\mathfrak{S}}} = \bigcup_{s_\alpha \in \mathfrak{S}, p_\alpha^2 \in P_{\underset{\sim}{A}_2^{\mathfrak{S}}}} \{\langle s_\alpha, p_\alpha^2\rangle\}$, some set operations on the PHFLTSs, which are also PHFLTSs, can be described as follows (see e.g. [1]):

$$h^c_{\underset{\sim}{A}^{\mathfrak{S}}} = \bigcup_{s_\alpha \in \mathfrak{S}, p_\alpha \in P_{\underset{\sim}{A}^{\mathfrak{S}}}} \{\langle s_\alpha, 1 - p_\alpha\rangle\}; \tag{2.23}$$

$$h_{\underset{\sim}{A}_1^{\mathfrak{S}}} \cup h_{\underset{\sim}{A}_2^{\mathfrak{S}}} = \bigcup_{s_\alpha \in \mathfrak{S}, p_\alpha^1 \in P_{\underset{\sim}{A}_1^{\mathfrak{S}}}, p_\alpha^2 \in P_{\underset{\sim}{A}_2^{\mathfrak{S}}}} \left\{\langle s_\alpha, \max\{p_\alpha^1, p_\alpha^2\}\rangle\right\}; \tag{2.24}$$

$$h_{\underset{\sim}{A}_1^{\mathfrak{S}}} \cap h_{\underset{\sim}{A}_2^{\mathfrak{S}}} = \bigcup_{s_\alpha \in \mathfrak{S}, p_\alpha^1 \in P_{\underset{\sim}{A}_1^{\mathfrak{S}}}, p_\alpha^2 \in P_{\underset{\sim}{A}_2^{\mathfrak{S}}}} \left\{\langle s_\alpha, \min\{p_\alpha^1, p_\alpha^2\}\rangle\right\}. \tag{2.25}$$

2.4 Hesitant Fuzzy Uncertain Linguistic Set

Lin et al. [10] introduced the concept of hesitant fuzzy uncertain linguistic set to overcome the limitation that arises from the case where the decision information about alternatives is usually uncertain due to the vagueness of inherent subjective nature of human think. In the sequel, Li et al. [6] presented

(continued)

a number of aggregation operators to deal with hesitant fuzzy uncertain linguistic multiple criteria decision making problems. Hou and Zhuo [4] investigated the multiple criteria decision making problems involving hesitant fuzzy uncertain linguistic information, and furthermore, they developed a number of aggregation operators of hesitant fuzzy uncertain linguistic information.

Definition 2.4 ([10]) Let X be a reference set, and $\mathfrak{S} = \{s_\alpha \mid \alpha = -\tau, \ldots, -1, 0, 1, \ldots, \tau\}$ be a linguistic term set. A hesitant fuzzy uncertain linguistic set (HFULS) on X is mathematically shown in terms of

$$\grave{A}^{\mathfrak{S}} = \{\langle x, h_{\grave{A}^{\mathfrak{S}}}(x)\rangle \mid x \in X\}. \tag{2.26}$$

Here, $h_{\grave{A}^{\mathfrak{S}}}(x)$ is a set of some possible pairs of interval-valued linguistic term set $\mathfrak{S}$ together with values in [0, 1] which can be characterized by

$$h_{\grave{A}^{\mathfrak{S}}}(x) = \left\langle [s_\alpha^L(x), s_\alpha^U(x)], \bigcup_{\gamma_{\grave{A}^{\mathfrak{S}}} \in h_{\grave{A}^{\mathfrak{S}}}(x)} \{\gamma_{\grave{A}^{\mathfrak{S}}}\} \right\rangle, \tag{2.27}$$

where $s_\alpha(x) \in \mathfrak{S}$ denotes the possible interval value for a linguistic variable involved in $h_{\grave{A}^{\mathfrak{S}}}(x)$.

Remark 2.4 Throughout this book, the set of all HFULSs on the reference set X is denoted by $\mathbb{HFULS}(X)$.

Example 2.3 Let $X = \{x_1, x_2\}$ be the reference set, $h_{\grave{A}^{\mathfrak{S}}}(x_1) = \langle [s_2, s_3], \{0.2, 0, 3, 0.6\}\rangle$ and $h_{\grave{A}^{\mathfrak{S}}}(x_2) = \langle [s_{-1}, s_0], \{0.3, 0.5, 0.7\}\rangle$ be the HFULEs of x_i $(i = 1, 2)$ to a set $\grave{A}^{\mathfrak{S}}$, respectively. Then $\grave{A}^{\mathfrak{S}}$ can be considered as an HFULS, i.e.,

$$\grave{A}^{\mathfrak{S}} = \{\langle x_1, \langle [s_2, s_3], \{0.2, 0, 3, 0.6\}\rangle\rangle, \langle x_2, \langle [s_{-1}, s_0], \{0.3, 0.5, 0.7\}\rangle\rangle\}.$$

Needless to say that for any three uncertain linguistic variables $s_\alpha = [s_\alpha^L, s_\alpha^U]$, $s_{\alpha_1} = [s_{\alpha_1}^L, s_{\alpha_1}^U]$, and $s_{\alpha_2} = [s_{\alpha_2}^L, s_{\alpha_2}^U]$, one concludes that (see, e.g., [10])

$$s_{\alpha_1} \oplus s_{\alpha_2} = \left[s_{\alpha_1}^L, s_{\alpha_1}^U\right] \oplus \left[s_{\alpha_2}^L, s_{\alpha_2}^U\right] = \left[s_{\alpha_1+\alpha_2}^L, s_{\alpha_1+\alpha_2}^U\right];$$

$$s_{\alpha_1} \otimes s_{\alpha_2} = \left[s_{\alpha_1}^L, s_{\alpha_1}^U\right] \otimes \left[s_{\alpha_2}^L, s_{\alpha_2}^U\right] = \left[s_{\alpha_1\times\alpha_2}^L, s_{\alpha_1\times\alpha_2}^U\right];$$

$$\lambda s_\alpha = \lambda\left[s_\alpha^L, s_\alpha^U\right] = \left[\lambda s_\alpha^L, \lambda s_\alpha^U\right] = \left[s_{\lambda\alpha}^L, s_{\lambda\alpha}^U\right], \quad \lambda > 0;$$

$$(s_\alpha)^\lambda = \left([s_\alpha^L, s_\alpha^U]\right)^\lambda = \left[(s_\alpha^L)^\lambda, (s_\alpha^U)^\lambda\right] = \left[s_{\alpha^\lambda}^L, s_{\alpha^\lambda}^U\right], \quad \lambda > 0.$$

Based on the above relationships between uncertain linguistic variables, we are able to define the following operations on the HFULSs $h_{\grave{A}^{\mathfrak{S}}} = \langle [s_\alpha^L, s_\alpha^U], \bigcup_{\gamma_{\grave{A}^{\mathfrak{S}}} \in h_{\grave{A}^{\mathfrak{S}}}} \{\gamma_{\grave{A}^{\mathfrak{S}}}\}\rangle$, $h_{\grave{A}_1^{\mathfrak{S}}} = \langle [s_{\alpha_1}^L, s_{\alpha_1}^U], \bigcup_{\gamma_{\grave{A}_1^{\mathfrak{S}}} \in h_{\grave{A}_1^{\mathfrak{S}}}} \{\gamma_{\grave{A}_1^{\mathfrak{S}}}\}\rangle$ and $h_{\grave{A}_2^{\mathfrak{S}}} = \langle [s_{\alpha_2}^L, s_{\alpha_2}^U], \bigcup_{\gamma_{\grave{A}_2^{\mathfrak{S}}} \in h_{\grave{A}_2^{\mathfrak{S}}}} \{\gamma_{\grave{A}_2^{\mathfrak{S}}}\}\rangle$ (see, e.g., [4]):

$$h_{\grave{A}_1^{\mathfrak{S}}} \oplus h_{\grave{A}_2^{\mathfrak{S}}} = \left\langle [s_{\alpha_1+\alpha_2}^L, s_{\alpha_1+\alpha_2}^U], \bigcup_{\gamma_{\grave{A}_1^{\mathfrak{S}}} \in h_{\grave{A}_1^{\mathfrak{S}}}, \gamma_{\grave{A}_2^{\mathfrak{S}}} \in h_{\grave{A}_2^{\mathfrak{S}}}} \{\gamma_{\grave{A}_1^{\mathfrak{S}}} + \gamma_{\grave{A}_2^{\mathfrak{S}}} - \gamma_{\grave{A}_1^{\mathfrak{S}}} \times \gamma_{\grave{A}_2^{\mathfrak{S}}}\}\right\rangle; \quad (2.28)$$

$$h_{\grave{A}_1^{\mathfrak{S}}} \otimes h_{\grave{A}_2^{\mathfrak{S}}} = \left\langle [s_{\alpha_1\times\alpha_2}^L, s_{\alpha_1\times\alpha_2}^U], \bigcup_{\gamma_{\grave{A}_1^{\mathfrak{S}}} \in h_{\grave{A}_1^{\mathfrak{S}}}, \gamma_{\grave{A}_2^{\mathfrak{S}}} \in h_{\grave{A}_2^{\mathfrak{S}}}} \{\gamma_{\grave{A}_1^{\mathfrak{S}}} \times \gamma_{\grave{A}_2^{\mathfrak{S}}}\}\right\rangle; \quad (2.29)$$

$${h_{\grave{A}^{\mathfrak{S}}}}^\lambda = \left\langle [s_{\alpha^\lambda}^L, s_{\alpha^\lambda}^U], \bigcup_{\gamma_{\grave{A}^{\mathfrak{S}}} \in h_{\grave{A}^{\mathfrak{S}}}} \{(\gamma_{\grave{A}^{\mathfrak{S}}})^\lambda\}\right\rangle, \quad \lambda > 0; \quad (2.30)$$

$$\lambda h_{\grave{A}^{\mathfrak{S}}} = \left\langle [s_{\lambda\alpha}^L, s_{\lambda\alpha}^U], \bigcup_{\gamma_{\grave{A}^{\mathfrak{S}}} \in h_{\grave{A}^{\mathfrak{S}}}} \{1 - (1 - \gamma_{\grave{A}^{\mathfrak{S}}})^\lambda\}\right\rangle, \quad \lambda > 0. \quad (2.31)$$

Remark 2.5 Some other relevant concepts including interval-valued hesitant fuzzy uncertain linguistic set (IVHFULS) [23], linguistic interval hesitant fuzzy set (LIHFS) [12], interval-valued 2-tuple hesitant fuzzy linguistic term set (IV2THFLTS) [16], and multi-hesitant fuzzy linguistic term set (MHFLTS) [19] can be defined similarly, and therefore, we do not show them here to save space.

2.5 Dual Hesitant Fuzzy Linguistic Set

Inspired by the idea of hesitant fuzzy linguistic variable, Yang and Ju [30] combined dual hesitant fuzzy set with linguistic term set to construct the concept of dual hesitant fuzzy linguistic set containing a linguistic term and a set of membership and non-membership degrees. Wei et al. [24] developed a number of aggregation operators with dual hesitant fuzzy linguistic information and utilized them in developing some approaches for dealing with hesitant fuzzy linguistic multiple criteria decision making problems. Zhang et al. [32] extended Archimedean t-norm and t-conorm in order to aggregate the dual hesitant fuzzy linguistic information. Li et al. [9] defined an extended TODIM method under dual hesitant fuzzy linguistic information, and then they applied that technique for dealing with a stock selection problem.

Definition 2.5 ([30]) Let X be a reference set, a dual hesitant fuzzy linguistic set (DHFLS) $\mathbf{A}^{\mathfrak{S}}$ on X is defined in terms of two functions $u_{\mathbf{A}^{\mathfrak{S}}}(x)$ and $v_{\mathbf{A}^{\mathfrak{S}}}(x)$ together with the linguistic term set $\mathfrak{S} = \{s_\alpha \mid \alpha = 0, 1, \ldots, \tau\}$ as follows:

$$\mathbf{A}^{\mathfrak{S}} = \{\langle x, s_\alpha(x), u_{\mathbf{A}^{\mathfrak{S}}}(x), v_{\mathbf{A}^{\mathfrak{S}}}(x)\rangle \mid x \in X\}, \tag{2.32}$$

where $u_{\mathbf{A}^{\mathfrak{S}}}(x)$ and $v_{\mathbf{A}^{\mathfrak{S}}}(x)$ are, respectively, the sets of some different values in $[0, 1]$ and represent the possible membership degrees and non-membership degrees of the element $x \in X$ with respect to the linguistic term set $\mathfrak{S}$.
Here, for all $x \in X$, if we take $u_{\mathbf{A}^{\mathfrak{S}}}(x) = \bigcup_{\gamma_{\mathbf{A}^{\mathfrak{S}}} \in u_{\mathbf{A}^{\mathfrak{S}}}(x)}\{\gamma_{\mathbf{A}^{\mathfrak{S}}}\}$, $v_{\mathbf{A}^{\mathfrak{S}}}(x) = \bigcup_{\eta_{\mathbf{A}^{\mathfrak{S}}} \in v_{\mathbf{A}^{\mathfrak{S}}}(x)}\{\eta_{\mathbf{A}^{\mathfrak{S}}}\}$, $\gamma^+_{\mathbf{A}^{\mathfrak{S}}} \in u^+_{\mathbf{A}^{\mathfrak{S}}} = \bigcup_{x \in X} \max_{\gamma_{\mathbf{A}^{\mathfrak{S}}} \in u_{\mathbf{A}^{\mathfrak{S}}}(x)}\{\gamma_{\mathbf{A}^{\mathfrak{S}}}\}$ and $\eta^+_{\mathbf{A}^{\mathfrak{S}}} \in v^+_{\mathbf{A}^{\mathfrak{S}}} = \bigcup_{x \in X} \max_{\eta_{\mathbf{A}^{\mathfrak{S}}} \in v_{\mathbf{A}^{\mathfrak{S}}}(x)}\{\eta_{\mathbf{A}^{\mathfrak{S}}}\}$, then we conclude that

$$0 \le \gamma_{\mathbf{A}^{\mathfrak{S}}},\ \eta_{\mathbf{A}^{\mathfrak{S}}} \le 1, \quad 0 \le \gamma^+_{\mathbf{A}^{\mathfrak{S}}} + \eta^+_{\mathbf{A}^{\mathfrak{S}}} \le 1.$$

For the sake of simplicity, $h_{\mathbf{A}^{\mathfrak{S}}}(x) = (s_\alpha(x), u_{\mathbf{A}^{\mathfrak{S}}}(x), v_{\mathbf{A}^{\mathfrak{S}}}(x))$ is called as the dual hesitant fuzzy linguistic element (DHFLE).

Remark 2.6 Throughout this book, the set of all DHFLSs on the reference set X is denoted by $\mathbb{DHFLS}(X)$.

Remark 2.7 It should be mentioned that the concept of DHFLS is also called in the literature [3] as the intuitionistic hesitant fuzzy linguistic set (IHFLS).

Example 2.4 Let $X = \{x_1, x_2\}$ be the reference set,
$h_{\mathbf{A}^{\mathfrak{S}}}(x_1) = ((s_\alpha(x_1), u_{\mathbf{A}^{\mathfrak{S}}}(x_1), v_{\mathbf{A}^{\mathfrak{S}}}(x_1)) = (s_2, \{0.2, 0.5\}, \{0.3\})$ and $h_{\mathbf{A}^{\mathfrak{S}}}(x_2) = (s_2(x_2), u_{\mathbf{A}^{\mathfrak{S}}}(x_2), v_{\mathbf{A}^{\mathfrak{S}}}(x_2)) = (s_3, \{0.3, 0.4\}, \{0.1, 0.6\})$ be the DHFLEs of x_i $(i = 1, 2)$ in the set $\mathbf{A}^{\mathfrak{S}}$, respectively. Then $\mathbf{A}^{\mathfrak{S}}$ can be considered as a DHFLS, i.e.,

$$\mathbf{A}^{\mathfrak{S}} = \{\langle x_1, s_2, \{0.2, 0.5\}, \{0.3\}\rangle, \langle x_2, s_3, \{0.3, 0.4\}, \{0.1, 0.6\}\rangle\}.$$

For the DHFLEs $h_{\mathbf{A}^{\mathfrak{S}}} = (s_\alpha, u_{\mathbf{A}^{\mathfrak{S}}}, v_{\mathbf{A}^{\mathfrak{S}}})$, $h_{\mathbf{A}_1^{\mathfrak{S}}} = (s_{\alpha_1}, u_{\mathbf{A}_1^{\mathfrak{S}}}, v_{\mathbf{A}_1^{\mathfrak{S}}})$, and $h_{\mathbf{A}_2^{\mathfrak{S}}} = (s_{\alpha_2}, u_{\mathbf{A}_2^{\mathfrak{S}}}, v_{\mathbf{A}_2^{\mathfrak{S}}})$ the following operations are defined (see [24]):

$$h_{\mathbf{A}_1^{\mathfrak{S}}} \oplus h_{\mathbf{A}_2^{\mathfrak{S}}} = (s_{\alpha_1+\alpha_2}, u_{\mathbf{A}_1^{\mathfrak{S}}} \oplus u_{\mathbf{A}_2^{\mathfrak{S}}}, v_{\mathbf{A}_1^{\mathfrak{S}}} \otimes v_{\mathbf{A}_2^{\mathfrak{S}}}); \tag{2.33}$$

$$h_{\mathbf{A}_1^{\mathfrak{S}}} \otimes h_{\mathbf{A}_2^{\mathfrak{S}}} = (s_{\alpha_1\times\alpha_2}, u_{\mathbf{A}_1^{\mathfrak{S}}} \otimes u_{\mathbf{A}_2^{\mathfrak{S}}}, v_{\mathbf{A}_1^{\mathfrak{S}}} \oplus v_{\mathbf{A}_2^{\mathfrak{S}}}), \tag{2.34}$$

$$\lambda h_{\mathbf{A}^{\mathfrak{S}}} = (s_{\lambda\alpha}, 1-(1-u_{\mathbf{A}^{\mathfrak{S}}})^\lambda, (v_{\mathbf{A}^{\mathfrak{S}}})^\lambda); \tag{2.35}$$

$$h_{\mathbf{A}^{\mathfrak{S}}}{}^\lambda = (s_{(\alpha)^\lambda}, (u_{\mathbf{A}^{\mathfrak{S}}})^\lambda, 1-(1-v_{\mathbf{A}^{\mathfrak{S}}})^\lambda), \tag{2.36}$$

where the operations $u_{\mathbf{A}_1^{\mathfrak{S}}} \circledast u_{\mathbf{A}_2^{\mathfrak{S}}}$ and $v_{\mathbf{A}_1^{\mathfrak{S}}} \circledast v_{\mathbf{A}_2^{\mathfrak{S}}}$ coincide with those defined between HFSs in the preceding sections.

On the basis of the above operations on DHFLEs, some relationships can be further established for the operations on DHFLSs as follows:

Let X be a reference set, $\mathbf{A}_1{}^{\mathfrak{S}}$ and $\mathbf{A}_2{}^{\mathfrak{S}}$ be two DHFLSs. We define

$$\begin{aligned}\mathbf{A}_1{}^{\mathfrak{S}} \oplus \mathbf{A}_2{}^{\mathfrak{S}} &= \bigcup_{h_{\mathbf{A}_1^{\mathfrak{S}}} \in \mathbf{A}_1{}^{\mathfrak{S}}, h_{\mathbf{A}_2^{\mathfrak{S}}} \in \mathbf{A}_2{}^{\mathfrak{S}}} h_{\mathbf{A}_1^{\mathfrak{S}}} \oplus h_{\mathbf{A}_2^{\mathfrak{S}}} \\ &= \{\langle x, s_{\alpha_1+\alpha_2}, u_{\mathbf{A}_1^{\mathfrak{S}}}(x) \oplus u_{\mathbf{A}_2^{\mathfrak{S}}}(x), \\ &\quad v_{\mathbf{A}_1^{\mathfrak{S}}}(x) \otimes v_{\mathbf{A}_2^{\mathfrak{S}}}(x)\rangle | x \in X\},\end{aligned} \tag{2.37}$$

$$\begin{aligned}\mathbf{A}_1{}^{\mathfrak{S}} \otimes \mathbf{A}_2{}^{\mathfrak{S}} &= \bigcup_{h_{\mathbf{A}_1^{\mathfrak{S}}} \in \mathbf{A}_1{}^{\mathfrak{S}}, h_{\mathbf{A}_2^{\mathfrak{S}}} \in \mathbf{A}_2{}^{\mathfrak{S}}} h_{\mathbf{A}_1^{\mathfrak{S}}} \otimes h_{\mathbf{A}_2^{\mathfrak{S}}} \\ &= \left\{\langle x, s_{\alpha_1\times\alpha_2}, u_{\mathbf{A}_1^{\mathfrak{S}}}(x) \otimes u_{\mathbf{A}_2^{\mathfrak{S}}}(x), \right. \\ &\quad v_{\mathbf{A}_1^{\mathfrak{S}}}(x) \oplus v_{\mathbf{A}_2^{\mathfrak{S}}}(x)\rangle | x \in X\}.\end{aligned} \tag{2.38}$$

2.6 Dual Hesitant Fuzzy Triangular Linguistic Set

On the basis of hesitant fuzzy linguistic set, triangular linguistic term set, and dual hesitant fuzzy set, Ju et al. [5] introduced the concept of dual hesitant fuzzy triangular linguistic set which is composed of a triangular linguistic term. Dual hesitant fuzzy triangular linguistic set is a set of membership and non-membership degrees for overcoming the shortcomings of hesitant fuzzy linguistic set.

Definition 2.6 ([5]) Let X be a reference set. A dual hesitant fuzzy triangular linguistic set (DHFTLS) $\breve{A}^{\mathfrak{S}}$ on X is defined in terms of two functions $u_{\breve{A}^{\mathfrak{S}}}(x)$

and $v_{\breve{A}^{\mathfrak{S}}}(x)$ together with the linguistic term set $\mathfrak{S} = \{s_\alpha \mid \alpha = 0, 1, \ldots, \tau\}$ as follows:

$$\breve{A}^{\mathfrak{S}} = \left\{ \langle x, \langle s_\alpha^L, s_\alpha^M, s_\alpha^U \rangle, u_{\breve{A}^{\mathfrak{S}}}(x), v_{\breve{A}^{\mathfrak{S}}}(x) \rangle \mid x \in X \right\}, \tag{2.39}$$

where $u_{\breve{A}^{\mathfrak{S}}}(x)$ and $v_{\breve{A}^{\mathfrak{S}}}(x)$ are the sets of some different values in $[0, 1]$ and represent the possible membership degrees and non-membership degrees of the element $x \in X$ to the triangular linguistic term set $\mathfrak{S}$, respectively.

Here, for all $x \in X$, if we denote $u_{\breve{A}^{\mathfrak{S}}}(x) = \bigcup_{\gamma_{\breve{A}^{\mathfrak{S}}} \in u_{\breve{A}^{\mathfrak{S}}}(x)} \{\gamma_{\breve{A}^{\mathfrak{S}}}\}$, $v_{\breve{A}^{\mathfrak{S}}}(x) = \bigcup_{\eta_{\breve{A}^{\mathfrak{S}}} \in v_{\breve{A}^{\mathfrak{S}}}(x)} \{\eta_{\breve{A}^{\mathfrak{S}}}\}$, $\gamma^+_{\breve{A}^{\mathfrak{S}}} \in u^+_{\breve{A}^{\mathfrak{S}}} = \bigcup_{x \in X} \max_{\gamma_{\breve{A}^{\mathfrak{S}}} \in u_{\breve{A}^{\mathfrak{S}}}(x)} \{\gamma_{\breve{A}^{\mathfrak{S}}}\}$ and $\eta^+_{\breve{A}^{\mathfrak{S}}} \in v^+_{\breve{A}^{\mathfrak{S}}} = \bigcup_{x \in X} \max_{\eta_{\breve{A}^{\mathfrak{S}}} \in v_{\breve{A}^{\mathfrak{S}}}(x)} \{\eta_{\breve{A}^{\mathfrak{S}}}\}$, then we conclude that

$$0 \le \gamma_{\breve{A}^{\mathfrak{S}}}, \ \eta_{\breve{A}^{\mathfrak{S}}} \le 1, \quad 0 \le \gamma^+_{\breve{A}^{\mathfrak{S}}} + \eta^+_{\breve{A}^{\mathfrak{S}}} \le 1.$$

For the sake of simplicity, $h_{\breve{A}^{\mathfrak{S}}}(x) = (\langle s_\alpha^L, s_\alpha^M, s_\alpha^U \rangle, u_{\breve{A}^{\mathfrak{S}}}(x), v_{\breve{A}^{\mathfrak{S}}}(x))$ is called as the dual hesitant fuzzy linguistic element (DHFTLE).

Remark 2.8 Throughout this book, the set of all DHFTLSs on the reference set X is denoted by $\mathbb{DHFTLS}(X)$.

Example 2.5 Let $X = \{x_1, x_2\}$ be the reference set, $h_{\breve{A}^{\mathfrak{S}}}(x_1) = (\langle s_\alpha^L, s_\alpha^M, s_\alpha^U \rangle, u_{\breve{A}^{\mathfrak{S}}}(x_1), v_{\breve{A}^{\mathfrak{S}}}(x_1)) = (\langle s_1, s_2, s_3 \rangle, \{0.2, 0.5\}, \{0.3\})$ and $h_{\breve{A}^{\mathfrak{S}}}(x_2) = (\langle s_\alpha^L, s_\alpha^M, s_\alpha^U \rangle, u_{\breve{A}^{\mathfrak{S}}}(x_2), v_{\breve{A}^{\mathfrak{S}}}(x_2)) = (\langle s_{-1}, s_0, s_1 \rangle, \{0.3, 0.4\}, \{0.1, 0.6\})$ be the DHFTLEs of x_i $(i = 1, 2)$ in the set $\breve{A}^{\mathfrak{S}}$, respectively. Then $\breve{A}^{\mathfrak{S}}$ can be considered as a DHFTLS, i.e.,

$$\breve{A}^{\mathfrak{S}} = \{\langle x_1, \langle s_1, s_2, s_3 \rangle, \{0.2, 0.5\}, \{0.3\} \rangle, \ \langle x_2, \langle s_{-1}, s_0, s_1 \rangle, \{0.3, 0.4\}, \{0.1, 0.6\} \rangle\}.$$

For the DHFTLEs $h_{\breve{A}^{\mathfrak{S}}} = (\langle s_\alpha^L, s_\alpha^M, s_\alpha^U \rangle, u_{\breve{A}^{\mathfrak{S}}}, v_{\breve{A}^{\mathfrak{S}}})$, $h_{\breve{A}_1^{\mathfrak{S}}} = (\langle s_{\alpha_1}^L, s_{\alpha_1}^M, s_{\alpha_1}^U \rangle, u_{\breve{A}_1^{\mathfrak{S}}}, v_{\breve{A}_1^{\mathfrak{S}}})$ and $h_{\breve{A}_2^{\mathfrak{S}}} = (\langle s_{\alpha_2}^L, s_{\alpha_2}^M, s_{\alpha_2}^U \rangle, u_{\breve{A}_2^{\mathfrak{S}}}, v_{\breve{A}_2^{\mathfrak{S}}})$ the following operations are defined (see [5]):

$$h_{\breve{A}_1^{\mathfrak{S}}} \oplus h_{\breve{A}_2^{\mathfrak{S}}} = \left(\langle s^L_{\alpha_1+\alpha_2}, s^M_{\alpha_1+\alpha_2}, s^U_{\alpha_1+\alpha_2} \rangle, u_{\breve{A}_1^{\mathfrak{S}}} \oplus u_{\breve{A}_2^{\mathfrak{S}}}, v_{\breve{A}_1^{\mathfrak{S}}} \otimes v_{\breve{A}_2^{\mathfrak{S}}} \right); \tag{2.40}$$

$$h_{\breve{A}_1^{\mathfrak{S}}} \otimes h_{\breve{A}_2^{\mathfrak{S}}} = \left(\langle s^L_{\alpha_1\times\alpha_2}, s^M_{\alpha_1\times\alpha_2}, s^U_{\alpha_1\times\alpha_2} \rangle, u_{\breve{A}_1^{\mathfrak{S}}} \otimes u_{\breve{A}_2^{\mathfrak{S}}}, v_{\breve{A}_1^{\mathfrak{S}}} \oplus v_{\breve{A}_2^{\mathfrak{S}}} \right), \tag{2.41}$$

$$\lambda h_{\breve{A}^{\mathfrak{S}}} = \left(\langle s^L_{\lambda\alpha}, s^M_{\lambda\alpha}, s^U_{\lambda\alpha} \rangle, 1 - (1 - u_{\breve{A}^{\mathfrak{S}}})^\lambda, (v_{\breve{A}^{\mathfrak{S}}})^\lambda \right); \tag{2.42}$$

$$h_{\breve{A}^{\mathfrak{S}}}{}^\lambda = \left(\langle s^L_{(\alpha)^\lambda}, s^M_{(\alpha)^\lambda}, s^U_{(\alpha)^\lambda} \rangle, (u_{\breve{A}^{\mathfrak{S}}})^\lambda, 1 - (1 - v_{\breve{A}^{\mathfrak{S}}})^\lambda \right), \tag{2.43}$$

where the operations $u_{\breve{A}_1^{\mathfrak{S}}} \circledast u_{\breve{A}_2^{\mathfrak{S}}}$ and $v_{\breve{A}_1^{\mathfrak{S}}} \circledast v_{\breve{A}_2^{\mathfrak{S}}}$ coincide with those defined between HFSs in the preceding sections.

On the basis of the above operations on DHFTLEs, some relationships can be further established for the operations on DHFTLSs as follows:

Definition 2.7 ([5]) Let X be a reference set, $\breve{A}_1^{\mathfrak{S}}$ and $\breve{A}_2^{\mathfrak{S}}$ be two DHFTLSs. We define

$$\begin{aligned}\breve{A}_1^{\mathfrak{S}} \oplus \breve{A}_2^{\mathfrak{S}} &= \bigcup_{h_{\breve{A}_1^{\mathfrak{S}}} \in \breve{A}_1^{\mathfrak{S}}, h_{\breve{A}_2^{\mathfrak{S}}} \in \breve{A}_2^{\mathfrak{S}}} h_{\breve{A}_1^{\mathfrak{S}}} \oplus h_{\breve{A}_2^{\mathfrak{S}}} \\ &= \Big\{\Big\langle x, \langle s^L_{\alpha_1+\alpha_2}, s^M_{\alpha_1+\alpha_2}, s^U_{\alpha_1+\alpha_2}\rangle, u_{\breve{A}_1^{\mathfrak{S}}}(x) \oplus u_{\breve{A}_2^{\mathfrak{S}}}(x), \\ &\qquad v_{\breve{A}_1^{\mathfrak{S}}}(x) \otimes v_{\breve{A}_2^{\mathfrak{S}}}(x)\Big\rangle \,|x \in X\Big\}, \end{aligned} \tag{2.44}$$

$$\begin{aligned}\breve{A}_1^{\mathfrak{S}} \otimes \breve{A}_2^{\mathfrak{S}} &= \bigcup_{h_{\breve{A}_1^{\mathfrak{S}}} \in \breve{A}_1^{\mathfrak{S}}, h_{\breve{A}_2^{\mathfrak{S}}} \in \breve{A}_2^{\mathfrak{S}}} h_{\breve{A}_1^{\mathfrak{S}}} \otimes h_{\breve{A}_2^{\mathfrak{S}}} \\ &= \Big\{\Big\langle x, \langle s^L_{\alpha_1\times\alpha_2}, s^M_{\alpha_1\times\alpha_2}, s^U_{\alpha_1\times\alpha_2}\rangle, u_{\breve{A}_1^{\mathfrak{S}}}(x) \otimes u_{\breve{A}_2^{\mathfrak{S}}}(x), \\ &\qquad v_{\breve{A}_1^{\mathfrak{S}}}(x) \oplus v_{\breve{A}_2^{\mathfrak{S}}}(x)\Big\rangle \,|x \in X\Big\}. \end{aligned} \tag{2.45}$$

2.7 Interval-Valued Dual Hesitant Fuzzy Linguistic Set

Interval-valued dual hesitant fuzzy linguistic set concept has been introduced for facilitating the calculation whenever we are required to consider linguistic evaluation sets by considering the interval-valued dual hesitant fuzzy sets [26]. Following this line of thought, Qi et al. [13] presented the concept of interval-valued dual hesitant fuzzy unbalanced linguistic set together with its power aggregation operators. Xian et al. [28] utilized the concept of generalized interval-valued intuitionistic fuzzy linguistic variable to define an induced hybrid operator for dealing with TOPSIS-based linguistic group decision making problems.

Definition 2.8 ([26]) Let X be a reference set. An interval-valued dual hesitant fuzzy linguistic set (IVDHFLS) $\widetilde{\mathbf{A}}^{\mathfrak{S}}$ on X is defined in terms of two functions

$u_{\widetilde{\mathbf{A}}^{\mathfrak{S}}}(x)$ and $v_{\widetilde{\mathbf{A}}^{\mathfrak{S}}}(x)$ together with the linguistic term set $\mathfrak{S} = \{s_\alpha \mid \alpha = 0, 1, \dots, \tau\}$ as follows:

$$\widetilde{\mathbf{A}}^{\mathfrak{S}} = \left\{ \langle x, [s_\alpha^L(x), s_\alpha^U(x)], u_{\widetilde{\mathbf{A}}^{\mathfrak{S}}}(x), v_{\widetilde{\mathbf{A}}^{\mathfrak{S}}}(x) \rangle \mid x \in X \right\}, \tag{2.46}$$

where $u_{\widetilde{\mathbf{A}}^{\mathfrak{S}}}(x) = \bigcup_{[\gamma^L_{\widetilde{\mathbf{A}}^{\mathfrak{S}}}, \gamma^U_{\widetilde{\mathbf{A}}^{\mathfrak{S}}}] \in u_{\widetilde{\mathbf{A}}^{\mathfrak{S}}}(x)} \{[\gamma^L_{\widetilde{\mathbf{A}}^{\mathfrak{S}}}, \gamma^U_{\widetilde{\mathbf{A}}^{\mathfrak{S}}}]\}$ and
$v_{\widetilde{\mathbf{A}}^{\mathfrak{S}}}(x) = \bigcup_{[\eta^L_{\widetilde{\mathbf{A}}^{\mathfrak{S}}}, \eta^U_{\widetilde{\mathbf{A}}^{\mathfrak{S}}}] \in v_{\widetilde{\mathbf{A}}^{\mathfrak{S}}}(x)} \{[\eta^L_{\widetilde{\mathbf{A}}^{\mathfrak{S}}}, \eta^U_{\widetilde{\mathbf{A}}^{\mathfrak{S}}}]\}$ are some different interval values in $[0, 1]$ and represent, respectively, the possible interval-valued membership degrees and interval-valued non-membership degrees of the element $x \in X$ to the continuous linguistic term set $\hat{\mathfrak{S}} = \{s_\alpha \mid s_0 \leq s_\alpha \leq s_\tau\}$.

Here, for all $x \in X$, if we consider $\gamma^{U+}_{\widetilde{\mathbf{A}}^{\mathfrak{S}}} \in u^U_{\widetilde{\mathbf{A}}^{\mathfrak{S}}} = \bigcup_{x \in X} \max\{\gamma^U_{\widetilde{\mathbf{A}}^{\mathfrak{S}}}(x)\}$ and $\eta^{U+}_{\widetilde{\mathbf{A}}^{\mathfrak{S}}} \in v^U_{\widetilde{\mathbf{A}}^{\mathfrak{S}}} = \bigcup_{x \in X} \max\{\eta^U_{\widetilde{\mathbf{A}}^{\mathfrak{S}}}(x)\}$, then we conclude that

$$0 \leq \gamma^L_{\widetilde{\mathbf{A}}^{\mathfrak{S}}}, \gamma^U_{\widetilde{\mathbf{A}}^{\mathfrak{S}}}, \eta^L_{\widetilde{\mathbf{A}}^{\mathfrak{S}}}, \eta^U_{\widetilde{\mathbf{A}}^{\mathfrak{S}}} \leq 1, \quad 0 \leq \gamma^{U+}_{\widetilde{\mathbf{A}}^{\mathfrak{S}}} + \eta^{U+}_{\widetilde{\mathbf{A}}^{\mathfrak{S}}} \leq 1.$$

For the sake of simplicity, $h_{\widetilde{\mathbf{A}}^{\mathfrak{S}}}(x) = ([s_\alpha^L(x), s_\alpha^U(x)], \bigcup_{[\gamma^L_{\widetilde{\mathbf{A}}^{\mathfrak{S}}}, \gamma^U_{\widetilde{\mathbf{A}}^{\mathfrak{S}}}] \in u_{\widetilde{\mathbf{A}}^{\mathfrak{S}}}(x)} \{[\gamma^L_{\widetilde{\mathbf{A}}^{\mathfrak{S}}}, \gamma^U_{\widetilde{\mathbf{A}}^{\mathfrak{S}}}]\},$ $\bigcup_{[\eta^L_{\widetilde{\mathbf{A}}^{\mathfrak{S}}}, \eta^U_{\widetilde{\mathbf{A}}^{\mathfrak{S}}}] \in v_{\widetilde{\mathbf{A}}^{\mathfrak{S}}}(x)} \{[\eta^L_{\widetilde{\mathbf{A}}^{\mathfrak{S}}}, \eta^U_{\widetilde{\mathbf{A}}^{\mathfrak{S}}}]\})$ is called the interval-valued dual hesitant fuzzy linguistic element (IVDHFLE).

Remark 2.9 Throughout this book, the set of all IVDHFLSs on the reference set X is denoted by $\mathbb{IVDHFLS}(X)$.

Remark 2.10 It should be mentioned that the concept of IVDHFLS is also called in the literature [28] as the interval-valued intuitionistic hesitant fuzzy linguistic set (IVIHFLS).

Example 2.6 Let $X = \{x_1, x_2\}$ be the reference set,
$h_{\widetilde{\mathbf{A}}^{\mathfrak{S}}}(x_1) = ([s_\alpha^L(x_1), s_\alpha^U(x_1)], \bigcup_{[\gamma^L_{\widetilde{\mathbf{A}}^{\mathfrak{S}}}, \gamma^U_{\widetilde{\mathbf{A}}^{\mathfrak{S}}}] \in u_{\widetilde{\mathbf{A}}^{\mathfrak{S}}}(x_1)} \{[\gamma^L_{\widetilde{\mathbf{A}}^{\mathfrak{S}}}, \gamma^U_{\widetilde{\mathbf{A}}^{\mathfrak{S}}}]\}$ $, \bigcup_{[\eta^L_{\widetilde{\mathbf{A}}^{\mathfrak{S}}}, \eta^U_{\widetilde{\mathbf{A}}^{\mathfrak{S}}}] \in v_{\widetilde{\mathbf{A}}^{\mathfrak{S}}}(x_1)} \{[\eta^L_{\widetilde{\mathbf{A}}^{\mathfrak{S}}}, \eta^U_{\widetilde{\mathbf{A}}^{\mathfrak{S}}}]\}) = ([s_1, s_3], [0.2, 0.5], [0.3; 0.3])$ and $h_{\widetilde{\mathbf{A}}^{\mathfrak{S}}}(x_2) = ([s_\alpha^L(x_2), s_\alpha^U(x_2)], \bigcup_{[\gamma^L_{\widetilde{\mathbf{A}}^{\mathfrak{S}}}, \gamma^U_{\widetilde{\mathbf{A}}^{\mathfrak{S}}}] \in u_{\widetilde{\mathbf{A}}^{\mathfrak{S}}}(x_2)} \{[\gamma^L_{\widetilde{\mathbf{A}}^{\mathfrak{S}}}, \gamma^U_{\widetilde{\mathbf{A}}^{\mathfrak{S}}}]\},$ $\bigcup_{[\eta^L_{\widetilde{\mathbf{A}}^{\mathfrak{S}}}, \eta^U_{\widetilde{\mathbf{A}}^{\mathfrak{S}}}] \in v_{\widetilde{\mathbf{A}}^{\mathfrak{S}}}(x_2)} \{[\eta^L_{\widetilde{\mathbf{A}}^{\mathfrak{S}}}, \eta^U_{\widetilde{\mathbf{A}}^{\mathfrak{S}}}]\}) = ([s_0, s_1], [0.3, 0.4], [0.1, 0.6])$ be the DHFTLEs of x_i $(i = 1, 2)$ in the set $\widetilde{\mathbf{A}}^{\mathfrak{S}}$, respectively. Then $\widetilde{\mathbf{A}}^{\mathfrak{S}}$ can be considered as a IVDHFLS, i.e.,

$$\widetilde{\mathbf{A}}^{\mathfrak{S}} = \{\langle x_1, [s_1, s_3], [0.2, 0.5], [0.3, 0.3] \rangle, \ \langle x_2, [s_0, s_1], [0.3, 0.4], [0.1, 0.6] \rangle\}.$$

In what follows and from [26], we are going to present some algebraic operations on the IVDHFLEs
$h_{\widetilde{\mathbf{A}}^{\mathfrak{S}}} = ([s_\alpha^L, s_\alpha^U], \bigcup_{[\gamma^L_{\widetilde{\mathbf{A}}^{\mathfrak{S}}}, \gamma^U_{\widetilde{\mathbf{A}}^{\mathfrak{S}}}] \in u_{\widetilde{\mathbf{A}}^{\mathfrak{S}}}} \{[\gamma^L_{\widetilde{\mathbf{A}}^{\mathfrak{S}}}, \gamma^U_{\widetilde{\mathbf{A}}^{\mathfrak{S}}}]\}, \bigcup_{[\eta^L_{\widetilde{\mathbf{A}}^{\mathfrak{S}}}, \eta^U_{\widetilde{\mathbf{A}}^{\mathfrak{S}}}] \in v_{\widetilde{\mathbf{A}}^{\mathfrak{S}}}} \{[\eta^L_{\widetilde{\mathbf{A}}^{\mathfrak{S}}}, \eta^U_{\widetilde{\mathbf{A}}^{\mathfrak{S}}}]\}),$

$$h_{\tilde{\mathbf{A}}_1^{\mathfrak{S}}} = ([s_{\alpha_1}^L, s_{\alpha_1}^U], \bigcup_{[\gamma_{\tilde{\mathbf{A}}_1^{\mathfrak{S}}}^L, \gamma_{\tilde{\mathbf{A}}_1^{\mathfrak{S}}}^U] \in u_{\tilde{\mathbf{A}}_1^{\mathfrak{S}}}} \{[\gamma_{\tilde{\mathbf{A}}_1^{\mathfrak{S}}}^L, \gamma_{\tilde{\mathbf{A}}_1^{\mathfrak{S}}}^U]\}, \bigcup_{[\eta_{\tilde{\mathbf{A}}_1^{\mathfrak{S}}}^L, \eta_{\tilde{\mathbf{A}}_1^{\mathfrak{S}}}^U] \in v_{\tilde{\mathbf{A}}_1^{\mathfrak{S}}}} \{[\eta_{\tilde{\mathbf{A}}_1^{\mathfrak{S}}}^L, \eta_{\tilde{\mathbf{A}}_1^{\mathfrak{S}}}^U]\})$$

and

$$h_{\tilde{\mathbf{A}}_2^{\mathfrak{S}}} = ([s_{\alpha_2}^L, s_{\alpha_2}^U], \bigcup_{[\gamma_{\tilde{\mathbf{A}}_2^{\mathfrak{S}}}^L, \gamma_{\tilde{\mathbf{A}}_2^{\mathfrak{S}}}^U] \in u_{\tilde{\mathbf{A}}_2^{\mathfrak{S}}}} \{[\gamma_{\tilde{\mathbf{A}}_2^{\mathfrak{S}}}^L, \gamma_{\tilde{\mathbf{A}}_2^{\mathfrak{S}}}^U]\}, \bigcup_{[\eta_{\tilde{\mathbf{A}}_2^{\mathfrak{S}}}^L, \eta_{\tilde{\mathbf{A}}_2^{\mathfrak{S}}}^U] \in v_{\tilde{\mathbf{A}}_2^{\mathfrak{S}}}} \{[\eta_{\tilde{\mathbf{A}}_2^{\mathfrak{S}}}^L, \eta_{\tilde{\mathbf{A}}_2^{\mathfrak{S}}}^U]\})$$

$$\begin{aligned} h_{\tilde{\mathbf{A}}_1^{\mathfrak{S}}} \oplus h_{\tilde{\mathbf{A}}_2^{\mathfrak{S}}} = ([s_{\alpha_1+\alpha_2}^L, s_{\alpha_1+\alpha_2}^U], \\ \bigcup_{\substack{[\gamma_{\tilde{\mathbf{A}}_1^{\mathfrak{S}}}^L, \gamma_{\tilde{\mathbf{A}}_1^{\mathfrak{S}}}^U] \in u_{\tilde{\mathbf{A}}_1^{\mathfrak{S}}} \\ [\gamma_{\tilde{\mathbf{A}}_2^{\mathfrak{S}}}^L, \gamma_{\tilde{\mathbf{A}}_2^{\mathfrak{S}}}^U] \in u_{\tilde{\mathbf{A}}_2^{\mathfrak{S}}}}} \left\{[\gamma_{\tilde{\mathbf{A}}_1^{\mathfrak{S}}}^L + \gamma_{\tilde{\mathbf{A}}_2^{\mathfrak{S}}}^L - \gamma_{\tilde{\mathbf{A}}_1^{\mathfrak{S}}}^L \times \gamma_{\tilde{\mathbf{A}}_2^{\mathfrak{S}}}^L, \gamma_{\tilde{\mathbf{A}}_1^{\mathfrak{S}}}^U + \gamma_{\tilde{\mathbf{A}}_2^{\mathfrak{S}}}^U - \gamma_{\tilde{\mathbf{A}}_1^{\mathfrak{S}}}^U \times \gamma_{\tilde{\mathbf{A}}_2^{\mathfrak{S}}}^U]\right\}, \\ \bigcup_{\substack{[\eta_{\tilde{\mathbf{A}}_1^{\mathfrak{S}}}^L, \eta_{\tilde{\mathbf{A}}_1^{\mathfrak{S}}}^U] \in v_{\tilde{\mathbf{A}}_1^{\mathfrak{S}}} \\ [\eta_{\tilde{\mathbf{A}}_2^{\mathfrak{S}}}^L, \eta_{\tilde{\mathbf{A}}_2^{\mathfrak{S}}}^U] \in v_{\tilde{\mathbf{A}}_2^{\mathfrak{S}}}}} \left\{[\eta_{\tilde{\mathbf{A}}_1^{\mathfrak{S}}}^L \times \eta_{\tilde{\mathbf{A}}_2^{\mathfrak{S}}}^L, \eta_{\tilde{\mathbf{A}}_1^{\mathfrak{S}}}^U \times \eta_{\tilde{\mathbf{A}}_2^{\mathfrak{S}}}^U]\right\}); \end{aligned} \tag{2.47}$$

$$\begin{aligned} h_{\tilde{\mathbf{A}}_1^{\mathfrak{S}}} \otimes h_{\tilde{\mathbf{A}}_2^{\mathfrak{S}}} = ([s_{\alpha_1\times\alpha_2}^L, s_{\alpha_1\times\alpha_2}^U], \\ \bigcup_{\substack{[\gamma_{\tilde{\mathbf{A}}_1^{\mathfrak{S}}}^L, \gamma_{\tilde{\mathbf{A}}_1^{\mathfrak{S}}}^U] \in u_{\tilde{\mathbf{A}}_1^{\mathfrak{S}}} \\ [\gamma_{\tilde{\mathbf{A}}_2^{\mathfrak{S}}}^L, \gamma_{\tilde{\mathbf{A}}_2^{\mathfrak{S}}}^U] \in u_{\tilde{\mathbf{A}}_2^{\mathfrak{S}}}}} \{[\gamma_{\tilde{\mathbf{A}}_1^{\mathfrak{S}}}^L \times \gamma_{\tilde{\mathbf{A}}_2^{\mathfrak{S}}}^L, \gamma_{\tilde{\mathbf{A}}_1^{\mathfrak{S}}}^U \times \gamma_{\tilde{\mathbf{A}}_2^{\mathfrak{S}}}^U]\}, \\ \bigcup_{\substack{[\eta_{\tilde{\mathbf{A}}_1^{\mathfrak{S}}}^L, \eta_{\tilde{\mathbf{A}}_1^{\mathfrak{S}}}^U] \in v_{\tilde{\mathbf{A}}_1^{\mathfrak{S}}} \\ [\eta_{\tilde{\mathbf{A}}_2^{\mathfrak{S}}}^L, \eta_{\tilde{\mathbf{A}}_2^{\mathfrak{S}}}^U] \in v_{\tilde{\mathbf{A}}_2^{\mathfrak{S}}}}} \Big\{[\eta_{\tilde{\mathbf{A}}_1^{\mathfrak{S}}}^L + \eta_{\tilde{\mathbf{A}}_2^{\mathfrak{S}}}^L - \eta_{\tilde{\mathbf{A}}_1^{\mathfrak{S}}}^L \times \eta_{\tilde{\mathbf{A}}_2^{\mathfrak{S}}}^L, \eta_{\tilde{\mathbf{A}}_1^{\mathfrak{S}}}^U \\ + \eta_{\tilde{\mathbf{A}}_2^{\mathfrak{S}}}^U - \eta_{\tilde{\mathbf{A}}_1^{\mathfrak{S}}}^U \times \eta_{\tilde{\mathbf{A}}_2^{\mathfrak{S}}}^U]\Big\}); \end{aligned} \tag{2.48}$$

$$\begin{aligned} \lambda h_{\tilde{\mathbf{A}}^{\mathfrak{S}}} = \Bigg([s_{\lambda\alpha}^L, s_{\lambda\alpha}^U], \\ \bigcup_{[\gamma_{\tilde{\mathbf{A}}^{\mathfrak{S}}}^L, \gamma_{\tilde{\mathbf{A}}^{\mathfrak{S}}}^U] \in u_{\tilde{\mathbf{A}}^{\mathfrak{S}}}} \left\{[1 - (1 - \gamma_{\tilde{\mathbf{A}}^{\mathfrak{S}}}^L)^\lambda, 1 - (1 - \gamma_{\tilde{\mathbf{A}}^{\mathfrak{S}}}^U)^\lambda]\right\}, \\ \bigcup_{[\eta_{\tilde{\mathbf{A}}^{\mathfrak{S}}}^L, \eta_{\tilde{\mathbf{A}}^{\mathfrak{S}}}^U] \in v_{\tilde{\mathbf{A}}^{\mathfrak{S}}}} \left\{[(\eta_{\tilde{\mathbf{A}}^{\mathfrak{S}}}^L)^\lambda, (\eta_{\tilde{\mathbf{A}}^{\mathfrak{S}}}^U)^\lambda]\right\}\Bigg); \end{aligned} \tag{2.49}$$

$$h_{\tilde{\mathbf{A}}^{\mathfrak{S}}}{}^{\lambda} = \Big([s_{(\alpha)^\lambda}^L, s_{(\alpha)^\lambda}^U],$$

$$\bigcup_{[\gamma^L_{\tilde{\mathbf{A}}\mathfrak{S}},\gamma^U_{\tilde{\mathbf{A}}\mathfrak{S}}]\in u_{\tilde{\mathbf{A}}\mathfrak{S}}} \left\{[(\gamma^L_{\tilde{\mathbf{A}}\mathfrak{S}})^\lambda, (\gamma^U_{\tilde{\mathbf{A}}\mathfrak{S}})^\lambda]\right\},$$

$$\left.\bigcup_{[\eta^L_{\tilde{\mathbf{A}}\mathfrak{S}},\eta^U_{\tilde{\mathbf{A}}\mathfrak{S}}]\in v_{\tilde{\mathbf{A}}\mathfrak{S}}} \left\{[1-(1-\eta^L_{\tilde{\mathbf{A}}\mathfrak{S}})^\lambda, 1-(1-\eta^U_{\tilde{\mathbf{A}}\mathfrak{S}})^\lambda]\right\}\right). \tag{2.50}$$

Remark 2.11 It should be mentioned that the concept of IVDHFLS is also called in the literature as the interval-valued dual hesitant fuzzy uncertain linguistic set (IVDHFULS) [22].

References

1. Z. Chen, K. Chin, Y. Li, Yi Yang, Proportional hesitant fuzzy linguistic term set for multiple criteria group decision making. Inf. Sci. **357**, 61–87 (2016)
2. S.M. Chen, J.A. Hong, Multicriteria linguistic decision making based on hesitant fuzzy linguistic term sets and the aggregation of fuzzy sets. Inf. Sci. **286**, 63–74 (2014)
3. M. Fanyong, T. Chunqiao, Distance measures for hesitant intuitionistic fuzzy linguistic sets. Econ. Comput. Econ. Cybern. Stud. Res. **51**, 1–18 (2017)
4. Z. Huo, Z. Zhou, Approaches to multiple attribute decision making with hesitant fuzzy uncertain linguistic information. J. Intell. Fuzzy Syst. **28**, 991–998 (2015)
5. Y. Ju, S. Yang, X. Liu, A novel method for multiattribute decision making with dual hesitant fuzzy triangular linguistic information. J. Appl. Math. **2014**, 909823 (2014)
6. Q. Li, X. Zhao, G. Wei, Model for software quality evaluation with hesitant fuzzy uncertain linguistic information. Int. J. Intell. Syst. **26**, 2639–2647 (2014)
7. H. Liao, Z. Xu, X.J. Zeng, J.M. Merigo, Qualitative decision making with correlation coefficients of hesitant fuzzy linguistic term sets. Knowl.-Based Syst. **76**, 127–138 (2015)
8. H.C. Liao, Z.S. Xu, X.J. Zeng, Distance and similarity measures for hesitant fuzzy linguistic term sets and their application in multi-criteria decision making. Inf. Sci. **271**, 125–142 (2014)
9. W. Li, Y. Zhuang, Z. Ren, An extended TODIM method and its application in the stock selection under dual hesitant fuzzy linguistic information. J. Intell. Fuzzy Syst. **37**, 7935–7950 (2019)
10. R. Lin, X. Zhao, G. Wei, Models for selecting an ERP system with hesitant fuzzy linguistic information. J. Intell. Fuzzy Syst. **26**, 2155–2165 (2014)
11. H.B. Liu, J.F. Cai, L. Jiang, On improving the additive consistency of the fuzzy preference relations based on comparative linguistic expressions. Int. J. Intell. Syst. **29**, 544–559 (2014)
12. F. Meng, X. Chen, Q. Zhang, Multi-attribute decision analysis under a linguistic hesitant fuzzy environment. Inf. Sci. **267**, 287–305 (2014)
13. X. Qi, J. Zhang, C. Liang, Multiple attributes group decision-making approaches based on interval-valued dual hesitant fuzzy unbalanced linguistic set and their applications. Complexity **2018**, 3172716 (2018)
14. R.M. Rodriguez, L. Martinez, F. Herrera, Hesitant fuzzy linguistic terms sets for decision making. IEEE Trans. Fuzzy Syst. **20**, 109–119 (2012)
15. R.M. Rodriguez, L. Martinez, F. Herrera, A group decision making model dealing with comparative linguistic expressions based on hesitant fuzzy linguistic term sets. Inf. Sci. **241**, 28–42 (2013)

16. G. Si, H. Liao, D. Yu, C. Llopis, Interval-valued 2-tuple hesitant fuzzy linguistic term set and its application in multiple attribute decision making. J. Intell. Fuzzy Syst. **34**, 4225–4236 (2018)
17. V. Torra, Y. Narukawa, On hesitant fuzzy sets and decision, in *The 18th IEEE International Conference on Fuzzy Systems, Jeju Island* (2009), pp. 1378–1382
18. H. Wang, Extended hesitant fuzzy linguistic term sets and their aggregation in group decision making. Int. J. Comput. Intell. Syst. **8**, 14–33 (2015)
19. J. Wang, J. Wang, H. Zhang, A likelihood-based TODIM approach based on multihesitant fuzzy linguistic information for evaluation in logistics outsourcing. Comput. Ind. Eng. **99**, 287–299 (2016)
20. J. Wang, J. Wu, J. Wang, H. Zhang, X. Chen, Interval-valued hesitant fuzzy linguistic sets and their applications in multi-criteria decision-making problems. Inf. Sci. **288**, 55–72 (2014)
21. J.H. Wang, J. Hao, A new version of 2-tuple fuzzy linguistic representation model for computing with words. IEEE Trans. Fuzzy Syst. **14**, 435–445 (2006)
22. G. Wei, Interval-valued dual hesitant fuzzy uncertain linguistic aggregation operators in multiple attribute decision making. J. Intell. Fuzzy Syst. **33**, 1881–1893 (2017)
23. G.W. Wei, Interval valued hesitant fuzzy uncertain linguistic aggregation operators in multiple attribute decision making. Int. J. Mach. Learn. Cybern. **7**, 1099–1114 (2016)
24. G. Wei, F. Alsaadi, T. Hayat, A. Alsaedi, Hesitant fuzzy linguistic arithmetic aggregation operators in multiple attribute decision making. Iranian J. Fuzzy Syst. **13**, 1–16 (2016)
25. C.P. Wei, N. Zhao, X.J. Tang, Operators and comparisons of hesitant fuzzy linguistic term sets. IEEE Trans. Fuzzy Syst. **22**, 575–585 (2014)
26. G. Wei, R. Lin, H. Wang, L. Ran, Interval-valued dual hesitant fuzzy linguistic arithmetic aggregation operators in multiple attribute decision making. Int. Core J. Eng. **1**, 2414–1895 (2015)
27. Z.B. Wu, J.P. Xu, Possibility distribution-based approach for MAGDM with hesitant fuzzy linguistic information. IEEE Trans. Cybern. **46**, 694–705 (2016)
28. S. Xian, Y. Dong, Y. Liu, N. Jing, A novel approach for linguistic group decision making based on generalized interval-valued intuitionistic fuzzy linguistic induced hybrid operator and TOPSIS. Int. J. Intell. Syst. **33**, 288–314 (2018)
29. Z.S. Xu, Deviation measures of linguistic preference relations in group decision making. Omega **33**, 249–254 (2005)
30. S. Yang, Y. Ju, Dual hesitant fuzzy linguistic aggregation operators and their applications to multi-attribute decision making. J. Intell. Fuzzy Syst. **27**, 1935–1947 (2014)
31. G.Q. Zhang, Y.C. Dong, Y.F. Xu, Consistency and consensus measures for linguistic preference relations based on distribution assessments. Inf. Fusion **17**, 46–55 (2014)
32. N. Zhang, Z. Yao, Y. Zhou, G. Wei, Some new dual hesitant fuzzy linguistic operators based on Archimedean t-norm and t-conorm. Neural Comput. Appl. **31**, 7017–7040 (2019)
33. B. Zhu, Z.S. Xu, Consistency measures for hesitant fuzzy linguistic preference relations. IEEE Trans. Fuzzy Syst. **22**, 35–45 (2014)

Chapter 3
Neutrosophic Hesitant Fuzzy Set

Abstract In this chapter, we first introduce briefly the concept of neutrosophic set, and then, we introduce the concept single-valued neutrosophic hesitant fuzzy set which is refereed here to as neutrosophic hesitant fuzzy set. In the sequel, we are going to argue about the concept of interval neutrosophic hesitant fuzzy set. Although, there may be reported some other kinds of neutrosophic hesitant fuzzy set by different researchers, the current two kinds of neutrosophic-based hesitant fuzzy sets are enough to fully understand the other kinds.

3.1 Neutrosophic Hesitant Fuzzy Set

Hesitant fuzzy set and neutrosophic set are two concepts which have been extended in two different directions so far. As a combination of these aforementioned concepts, Ye [12] introduced the concept of single-valued neutrosophic hesitant fuzzy set, which is also known as neutrosophic hesitant fuzzy set. Single-valued neutrosophic hesitant fuzzy set not only considers the truth, indeterminacy, and falsity membership degrees of neutrosophic set into account but also it extends these membership degrees to a set of possible values in a unique interval and easily expresses the incomplete and inconsistent information. However, in the sequel, Liu and Luo [6] presented the certainty, score, and accuracy functions of single-valued neutrosophic hesitant fuzzy set together with proposing the single-valued neutrosophic hesitant fuzzy ordered weighted averaging operator. Sahin and Liu [8] investigated some decision making problems in which the correlation coefficient of single-valued neutrosophic hesitant fuzzy information plays the key role. Li and Zhang [4] increased the potential of multiple criteria decision making by presenting the Choquet aggregation operator for single-valued neutrosophic hesitant fuzzy information. Some other kinds of aggregation operators of single-valued neutrosophic hesitant fuzzy sets were developed by Juan-Juan et al. [2], and Wang and Li [10].

B. Farhadinia, *Hesitant Fuzzy Set*, Computational Intelligence Methods and Applications, https://doi.org/10.1007/978-981-16-7301-6_3

By taking the reference set X into account, a neutrosophic set (NS) A on X is defined in terms of three functions $u_A(x)$, $w_A(x)$, and $v_A(x)$ as follows (see [7]):

$$A = \{\langle x, u_A(x), w_A(x), v_A(x)\rangle \mid x \in X\}, \tag{3.1}$$

where $u_A(x), w_A(x), v_A(x) \in [0, 1]$ are, respectively, called truth, indeterminacy, and falsity membership degrees of the element $x \in X$ to A.
Generally, each term $u_A(x)$, $w_A(x)$, or $v_A(x)$ may be real standard or non-standard subsets belong to $]0^-, 1^+[$.

Remark 3.1 In the case where the terms $u_A(x)$, $w_A(x)$, and $v_A(x)$ belong to $[0, 1]$, then the NS A is called single-valued neutrosophic set (SVNS).

Definition 3.1 ([12]) Let X be a reference set. A single-valued neutrosophic hesitant fuzzy set (SVNHFS) A on X is defined in terms of three functions $u_A(x)$, $w_A(x)$, and $v_A(x)$ as follows:

$$A = \{\langle x, u_A(x), w_A(x), v_A(x)\rangle \mid x \in X\}, \tag{3.2}$$

where $u_A(x), w_A(x), v_A(x) \in [0, 1]$ denote the possible truth, indeterminacy, and falsity hesitant membership degrees of the element $x \in X$ to A, respectively.
In this regard, a SVNHFS may be presented by the form of

$$A = \left\{ \langle x, \bigcup_{\gamma_A \in u_A} \{\gamma_A\}, \bigcup_{\delta_A \in w_A} \{\delta_A\}, \bigcup_{\eta_A \in v_A} \{\eta_A\}\rangle \mid x \in X \right\}. \tag{3.3}$$

For the sake of simplicity, $h_A(x) = (u_A(x), w_A(x), v_A(x))$ is called as the single-valued neutrosophic hesitant fuzzy element (SVNHFE).

Remark 3.2 Throughout this book, the set of all SVNHFSs on the reference set X is denoted by $\mathbb{SVNHFS}(X)$.

Example 3.1 Let $X = \{x_1, x_2\}$ be the reference set,
$h_A(x_1) = \{(\{0.2, 0, 3\}, \{0.4, 0.6\}, \{0.3\})\}$ and $h_A(x_2) = \{(\{0.3\}, \{0.5\}, \{0.6, 0.7, 0.8\})\}$ be the SVNHFEs of x_i $(i = 1, 2)$ to a set A, respectively. Then A can be considered as an SVNHFS, i.e.,

$$A = \{\langle x_1, (\{0.2, 0, 3\}, \{0.4, 0.6\}, \{0.3\})\rangle, \langle x_2, (\{0.3\}, \{0.5\}, \{0.6, 0.7, 0.8\})\rangle\}.$$

Given three SVNHFEs represented by
$h_A = (\bigcup_{\gamma_A \in u_A}\{\gamma_A\}, \bigcup_{\delta_A \in w_A}\{\delta_A\}, \bigcup_{\eta_A \in v_A}\{\eta_A\})$,
$h_{A_1} = (\bigcup_{\gamma_{A_1} \in u_{A_1}}\{\gamma_{A_1}\}, \bigcup_{\delta_{A_1} \in w_{A_1}}\{\delta_{A_1}\}, \bigcup_{\eta_{A_1} \in v_{A_1}}\{\eta_{A_1}\})$,
$h_{A_2} = (\bigcup_{\gamma_{A_2} \in u_{A_2}}\{\gamma_{A_2}\}, \bigcup_{\delta_{A_2} \in w_{A_2}}\{\delta_{A_2}\}, \bigcup_{\eta_{A_2} \in v_{A_2}}\{\eta_{A_2}\})$, a number of set and algebraic operations on the SVNHFEs, which are also SVNHFEs, can be described

as follows (see, e.g., [1]):

$$h_{A_1} \cup h_{A_2} = \left(\bigcup_{\substack{\gamma_{A_1} \in u_{A_1} \\ \gamma_{A_2} \in u_{A_2}}} \max\{\gamma_{A_1}, \gamma_{A_2}\}, \bigcup_{\substack{\delta_{A_1} \in w_{A_1} \\ \delta_{A_2} \in w_{A_2}}} \min\{\delta_{A_1}, \delta_{A_2}\}, \bigcup_{\substack{\eta_{A_1} \in v_{A_1} \\ \eta_{A_2} \in v_{A_2}}} \min\{\eta_{A_1}, \eta_{A_2}\} \right); \tag{3.4}$$

$$h_{A_1} \cap h_{A_2} = \left(\bigcup_{\substack{\gamma_{A_1} \in u_{A_1} \\ \gamma_{A_2} \in u_{A_2}}} \min\{\gamma_{A_1}, \gamma_{A_2}\}, \bigcup_{\substack{\delta_{A_1} \in w_{A_1} \\ \delta_{A_2} \in w_{A_2}}} \max\{\delta_{A_1}, \delta_{A_2}\}, \bigcup_{\substack{\eta_{A_1} \in v_{A_1} \\ \eta_{A_2} \in v_{A_2}}} \max\{\eta_{A_1}, \eta_{A_2}\} \right); \tag{3.5}$$

$$h_{A_1} \oplus h_{A_2} = \left(\bigcup_{\substack{\gamma_{A_1} \in u_{A_1} \\ \gamma_{A_2} \in u_{A_2}}} \{\gamma_{A_1} + \gamma_{A_2} - \gamma_{A_1} \times \gamma_{A_2}\}, \bigcup_{\substack{\delta_{A_1} \in w_{A_1} \\ \delta_{A_2} \in w_{A_2}}} \{\delta_{A_1} \times \delta_{A_2}\}, \bigcup_{\substack{\eta_{A_1} \in v_{A_1} \\ \eta_{A_2} \in v_{A_2}}} \{\eta_{A_1} \times \eta_{A_2}\} \right); \tag{3.6}$$

$$h_{A_1} \otimes h_{A_2} = \left(\bigcup_{\substack{\gamma_{A_1} \in u_{A_1} \\ \gamma_{A_2} \in u_{A_2}}} \{\gamma_{A_1} \times \gamma_{A_2}\}, \bigcup_{\substack{\delta_{A_1} \in w_{A_1} \\ \delta_{A_2} \in w_{A_2}}} \{\delta_{A_1} + \delta_{A_2} - \delta_{A_1} \times \delta_{A_2}\}, \bigcup_{\substack{\eta_{A_1} \in v_{A_1} \\ \eta_{A_2} \in v_{A_2}}} \{\eta_{A_1} + \eta_{A_2} - \eta_{A_1} \times \eta_{A_2}\} \right); \tag{3.7}$$

$$\lambda h_A = \left(\bigcup_{\gamma_A \in u_A} \{1 - (1 - \gamma_A)^{\lambda}\}, \bigcup_{\delta_A \in w_A} \{\delta_A^{\lambda}\}, \bigcup_{\eta_A \in v_A} \{\eta_A^{\lambda}\} \right); \tag{3.8}$$

$$h_A^\lambda = \left(\bigcup_{\gamma_A \in u_A} \{\gamma_A^\lambda\}, \bigcup_{\delta_A \in w_A} \{1-(1-\delta_A)^\lambda\}, \bigcup_{\eta_A \in v_A} \{1-(1-\eta_A)^\lambda\} \right), \tag{3.9}$$

where $\lambda > 0$.

3.2 Interval Neutrosophic Hesitant Fuzzy Set

Liu and Shi [5] introduced the concept of interval neutrosophic hesitant fuzzy set by the use of combination of interval-valued hesitant fuzzy set and interval neutrosophic set, which extends the truth, indeterminacy, and falsity membership functions of an element to a given set to interval-valued hesitant fuzzy set. Kakati et al. [3] represented the interval neutrosophic hesitant fuzzy Einstein Choquet integral operator, and they discussed thoroughly its properties. Ye [13] developed correlation coefficients of interval neutrosophic hesitant fuzzy sets, and he investigates the relation between correlation coefficients and similarity measures.

Definition 3.2 ([5]) Let X be a reference set, an interval neutrosophic hesitant fuzzy set (INHFS) $\widetilde{A}$ on X is defined in terms of three functions $u_{\widetilde{A}}(x)$, $w_{\widetilde{A}}(x)$, and $v_{\widetilde{A}}(x)$ as follows:

$$\widetilde{A} = \{\langle x, u_{\widetilde{A}}(x), w_{\widetilde{A}}(x), v_{\widetilde{A}}(x)\rangle \mid x \in X\}, \tag{3.10}$$

where $u_{\widetilde{A}}(x) = \bigcup_{[\gamma_{\widetilde{A}}^L, \gamma_{\widetilde{A}}^U] \in u_{\widetilde{A}}(x)} \{[\gamma_{\widetilde{A}}^L, \gamma_{\widetilde{A}}^U]\}$, $w_{\widetilde{A}}(x) = \bigcup_{[\delta_{\widetilde{A}}^L, \delta_{\widetilde{A}}^U] \in w_{\widetilde{A}}(x)} \{[\delta_{\widetilde{A}}^L, \delta_{\widetilde{A}}^U]\}$, and $v_{\widetilde{A}}(x) = \bigcup_{[\eta_{\widetilde{A}}^L, \eta_{\widetilde{A}}^U] \in v_{\widetilde{A}}(x)} \{[\eta_{\widetilde{A}}^L, \eta_{\widetilde{A}}^U]\}$ all are interval values in [0, 1], and denote the possible truth, indeterminacy, and falsity hesitant membership degrees of the element $x \in X$ to $\widetilde{A}$, respectively.

In this regard, an INHFS may be presented by the form of

$$\widetilde{A} = \left\{ \langle x, \bigcup_{[\gamma_{\widetilde{A}}^L, \gamma_{\widetilde{A}}^U] \in u_{\widetilde{A}}(x)} \{[\gamma_{\widetilde{A}}^L, \gamma_{\widetilde{A}}^U]\}, \bigcup_{[\delta_{\widetilde{A}}^L, \delta_{\widetilde{A}}^U] \in w_{\widetilde{A}}(x)} \{[\delta_{\widetilde{A}}^L, \delta_{\widetilde{A}}^U]\}, \bigcup_{[\eta_{\widetilde{A}}^L, \eta_{\widetilde{A}}^U] \in v_{\widetilde{A}}(x)} \{[\eta_{\widetilde{A}}^L, \eta_{\widetilde{A}}^U]\}\rangle \mid x \in X \right\}. \tag{3.11}$$

For the sake of simplicity, $h_{\widetilde{A}}(x) = (u_{\widetilde{A}}(x), w_{\widetilde{A}}(x), v_{\widetilde{A}}(x)) = (\bigcup_{[\gamma^L_{\widetilde{A}}, \gamma^U_{\widetilde{A}}] \in u_{\widetilde{A}}(x)} \{[\gamma^L_{\widetilde{A}}, \gamma^U_{\widetilde{A}}]\}, \bigcup_{[\delta^L_{\widetilde{A}}, \delta^U_{\widetilde{A}}] \in w_{\widetilde{A}}(x)} \{[\delta^L_{\widetilde{A}}, \delta^U_{\widetilde{A}}]\}, \bigcup_{[\eta^L_{\widetilde{A}}, \eta^U_{\widetilde{A}}] \in v_{\widetilde{A}}(x)} \{[\eta^L_{\widetilde{A}}, \eta^U_{\widetilde{A}}]\})$ is called as the interval neutrosophic hesitant fuzzy element (INHFE).

Remark 3.3 Throughout this book, the set of all INHFSs on the reference set X is denoted by $\mathbb{INHFS}(X)$.

Example 3.2 Let $X = \{x_1, x_2\}$ be the reference set, $h_{\widetilde{A}}(x_1) = \{(\{[0.2, 0, 3], [0.4, 0.5]\}, \{[0.4, 0.6]\}, \{[0.4, 0.5]\})\}$ and $h_{\widetilde{A}}(x_2) = \{(\{[0.3, 0.3]\}, \{[0.5, 0.6], [0.6, 0.7]\}, \{[0.6, 0.8]\})\}$ be the SVNHFEs of x_i $(i = 1, 2)$ to a set $\widetilde{A}$, respectively. Then $\widetilde{A}$ can be considered as an INHFS, i.e.,

$$\begin{aligned}\widetilde{A} = \{&\langle x_1, (\{[0.2, 0, 3], [0.4, 0.5]\}, \{[0.4, 0.6]\}, \{[0.4, 0.5]\}),\\ &\langle x_2, (\{[0.3, 0.3]\}, \{[0.5, 0.6], [0.6, 0.7]\}, \{[0.6, 0.8]\})\rangle\}.\end{aligned}$$

Given three INHFEs represented by $h_{\widetilde{A}} = (\bigcup_{[\gamma^L_{\widetilde{A}}, \gamma^U_{\widetilde{A}}] \in u_{\widetilde{A}}} \{[\gamma^L_{\widetilde{A}}, \gamma^U_{\widetilde{A}}]\}, \bigcup_{[\delta^L_{\widetilde{A}}, \delta^U_{\widetilde{A}}] \in w_{\widetilde{A}}} \{[\delta^L_{\widetilde{A}}, \delta^U_{\widetilde{A}}]\}, \bigcup_{[\eta^L_{\widetilde{A}}, \eta^U_{\widetilde{A}}] \in v_{\widetilde{A}}} \{[\eta^L_{\widetilde{A}}, \eta^U_{\widetilde{A}}]\})$, $h_{\widetilde{A}_1} = (\bigcup_{[\gamma^L_{\widetilde{A}_1}, \gamma^U_{\widetilde{A}_1}] \in u_{\widetilde{A}_1}} \{[\gamma^L_{\widetilde{A}_1}, \gamma^U_{\widetilde{A}_1}]\}, \bigcup_{[\delta^L_{\widetilde{A}_1}, \delta^U_{\widetilde{A}_1}] \in w_{\widetilde{A}_1}} \{[\delta^L_{\widetilde{A}_1}, \delta^U_{\widetilde{A}_1}]\}, \bigcup_{[\eta^L_{\widetilde{A}_1}, \eta^U_{\widetilde{A}_1}] \in v_{\widetilde{A}_1}} \{[\eta^L_{\widetilde{A}_1}, \eta^U_{\widetilde{A}_1}]\})$ and $h_{\widetilde{A}_2} = (\bigcup_{[\gamma^L_{\widetilde{A}_2}, \gamma^U_{\widetilde{A}_2}] \in u_{\widetilde{A}_2}} \{[\gamma^L_{\widetilde{A}_2}, \gamma^U_{\widetilde{A}_2}]\}, \bigcup_{[\delta^L_{\widetilde{A}_2}, \delta^U_{\widetilde{A}_2}] \in w_{\widetilde{A}_2}} \{[\delta^L_{\widetilde{A}_2}, \delta^U_{\widetilde{A}_2}]\}, \bigcup_{[\eta^L_{\widetilde{A}_2}, \eta^U_{\widetilde{A}_2}] \in v_{\widetilde{A}_2}} \{[\eta^L_{\widetilde{A}_2}, \eta^U_{\widetilde{A}_2}]\})$, some set and algebraic operations on the INHFEs, which are also INHFEs, can be described as follows (see, e.g., [5]):

$$\begin{aligned} h_{\widetilde{A}_1} \cup h_{\widetilde{A}_2} = \Bigg(&\bigcup_{\substack{[\gamma^L_{\widetilde{A}_1}, \gamma^U_{\widetilde{A}_1}] \in u_{\widetilde{A}_1} \\ [\gamma^L_{\widetilde{A}_2}, \gamma^U_{\widetilde{A}_2}] \in u_{\widetilde{A}_2}}} \left[\max\{\gamma^L_{\widetilde{A}_1}, \gamma^L_{\widetilde{A}_2}\}, \max\{\gamma^U_{\widetilde{A}_1}, \gamma^U_{\widetilde{A}_2}\}\right], \\ &\bigcup_{\substack{[\delta^L_{\widetilde{A}_1}, \delta^U_{\widetilde{A}_1}] \in w_{\widetilde{A}_1} \\ [\delta^L_{\widetilde{A}_2}, \delta^U_{\widetilde{A}_2}] \in w_{\widetilde{A}_2}}} \left[\min\{\delta^L_{\widetilde{A}_1}, \delta^L_{\widetilde{A}_2}\}, \min\{\delta^U_{\widetilde{A}_1}, \delta^U_{\widetilde{A}_2}\}\right], \\ &\bigcup_{\substack{[\eta^L_{\widetilde{A}_1}, \eta^U_{\widetilde{A}_1}] \in v_{\widetilde{A}_1} \\ [\eta^L_{\widetilde{A}_2}, \eta^U_{\widetilde{A}_2}] \in v_{\widetilde{A}_2}}} \left[\min\{\eta^L_{\widetilde{A}_1}, \eta^L_{\widetilde{A}_2}\}, \min\{\eta^U_{\widetilde{A}_1}, \eta^U_{\widetilde{A}_2}\}\right] \Bigg); \end{aligned} \tag{3.12}$$

$$h_{\widetilde{A}_1} \cap h_{\widetilde{A}_2} = \left(\bigcup_{\substack{[\gamma^L_{\widetilde{A}_1}, \gamma^U_{\widetilde{A}_1}] \in u_{\widetilde{A}_1} \\ [\gamma^L_{\widetilde{A}_2}, \gamma^U_{\widetilde{A}_2}] \in u_{\widetilde{A}_2}}} \left[\min\{\gamma^L_{\widetilde{A}_1}, \gamma^L_{\widetilde{A}_2}\}, \min\{\gamma^U_{\widetilde{A}_1}, \gamma^U_{\widetilde{A}_2}\} \right], \bigcup_{\substack{[\delta^L_{\widetilde{A}_1}, \delta^U_{\widetilde{A}_1}] \in w_{\widetilde{A}_1} \\ [\delta^L_{\widetilde{A}_2}, \delta^U_{\widetilde{A}_2}] \in w_{\widetilde{A}_2}}} \left[\max\{\delta^L_{\widetilde{A}_1}, \delta^L_{\widetilde{A}_2}\}, \max\{\delta^U_{\widetilde{A}_1}, \delta^U_{\widetilde{A}_2}\} \right], \bigcup_{\substack{[\eta^L_{\widetilde{A}_1}, \eta^U_{\widetilde{A}_1}] \in v_{\widetilde{A}_1} \\ [\eta^L_{\widetilde{A}_2}, \eta^U_{\widetilde{A}_2}] \in v_{\widetilde{A}_2}}} \left[\max\{\eta^L_{\widetilde{A}_1}, \eta^L_{\widetilde{A}_2}\}, \max\{\eta^U_{\widetilde{A}_1}, \eta^U_{\widetilde{A}_2}\} \right] \right); \tag{3.13}$$

$$h_{\widetilde{A}_1} \oplus h_{\widetilde{A}_2} = \left(\bigcup_{\substack{[\gamma^L_{\widetilde{A}_1}, \gamma^U_{\widetilde{A}_1}] \in u_{\widetilde{A}_1} \\ [\gamma^L_{\widetilde{A}_2}, \gamma^U_{\widetilde{A}_2}] \in u_{\widetilde{A}_2}}} \left\{ [\gamma^L_{\widetilde{A}_1} + \gamma^L_{\widetilde{A}_2} - \gamma^L_{\widetilde{A}_1} \times \gamma^L_{\widetilde{A}_2}, \gamma^U_{\widetilde{A}_1} + \gamma^U_{\widetilde{A}_2} - \gamma^U_{\widetilde{A}_1} \times \gamma^U_{\widetilde{A}_2}] \right\}, \bigcup_{\substack{[\delta^L_{\widetilde{A}_1}, \delta^U_{\widetilde{A}_1}] \in w_{\widetilde{A}_1} \\ [\delta^L_{\widetilde{A}_2}, \delta^U_{\widetilde{A}_2}] \in w_{\widetilde{A}_2}}} \left\{ [\delta^L_{\widetilde{A}_1} \times \delta^L_{\widetilde{A}_2}, \delta^U_{\widetilde{A}_1} \times \delta^U_{\widetilde{A}_2}] \right\}, \bigcup_{\substack{[\eta^L_{\widetilde{A}_1}, \eta^U_{\widetilde{A}_1}] \in v_{\widetilde{A}_1} \\ [\eta^L_{\widetilde{A}_2}, \eta^U_{\widetilde{A}_2}] \in v_{\widetilde{A}_2}}} \left\{ [\eta^L_{\widetilde{A}_1} \times \eta^L_{\widetilde{A}_2}, \eta^U_{\widetilde{A}_1} \times \eta^U_{\widetilde{A}_2}] \right\} \right); \tag{3.14}$$

$$h_{\widetilde{A}_1} \otimes h_{\widetilde{A}_2} = \left(\bigcup_{\substack{[\gamma^L_{\widetilde{A}_1}, \gamma^U_{\widetilde{A}_1}] \in u_{\widetilde{A}_1} \\ [\gamma^L_{\widetilde{A}_2}, \gamma^U_{\widetilde{A}_2}] \in u_{\widetilde{A}_2}}} \left\{ [\gamma^L_{\widetilde{A}_1} \times \gamma^L_{\widetilde{A}_2}, \gamma^U_{\widetilde{A}_1} \times \gamma^U_{\widetilde{A}_2}] \right\}, \bigcup_{\substack{[\delta^L_{\widetilde{A}_1}, \delta^U_{\widetilde{A}_1}] \in w_{\widetilde{A}_1} \\ [\delta^L_{\widetilde{A}_2}, \delta^U_{\widetilde{A}_2}] \in w_{\widetilde{A}_2}}} \left\{ [\delta^L_{\widetilde{A}_1} + \delta^L_{\widetilde{A}_2} - \delta^L_{\widetilde{A}_1} \times \delta^L_{\widetilde{A}_2}, \delta^U_{\widetilde{A}_1} + \delta^U_{\widetilde{A}_2} - \delta^U_{\widetilde{A}_1} \times \delta^U_{\widetilde{A}_2}] \right\}, \bigcup_{\substack{[\eta^L_{\widetilde{A}_1}, \eta^U_{\widetilde{A}_1}] \in v_{\widetilde{A}_1} \\ [\eta^L_{\widetilde{A}_2}, \eta^U_{\widetilde{A}_2}] \in v_{\widetilde{A}_2}}} \left\{ [\eta^L_{\widetilde{A}_1} + \eta^L_{\widetilde{A}_2} - \eta^L_{\widetilde{A}_1} \times \eta^L_{\widetilde{A}_2}, \eta^U_{\widetilde{A}_1} + \eta^U_{\widetilde{A}_2} - \eta^U_{\widetilde{A}_1} \times \eta^U_{\widetilde{A}_2}] \right\} \right); \tag{3.15}$$

$$\lambda h_{\widetilde{A}} = \left(\bigcup_{[\gamma^L_{\widetilde{A}}, \gamma^U_{\widetilde{A}}] \in u_{\widetilde{A}}} \{[1 - (1 - \gamma^L_{\widetilde{A}})^\lambda, 1 - (1 - \gamma^U_{\widetilde{A}})^\lambda]\}, \bigcup_{[\delta^L_{\widetilde{A}}, \delta^U_{\widetilde{A}}] \in w_{\widetilde{A}}} \left\{ [(\delta^L_{\widetilde{A}})^\lambda, (\delta^U_{\widetilde{A}})^\lambda] \right\}, \bigcup_{[\eta^L_{\widetilde{A}}, \eta^U_{\widetilde{A}}] \in v_{\widetilde{A}}} \left\{ [(\eta^L_{\widetilde{A}})^\lambda, (\eta^U_{\widetilde{A}})^\lambda] \right\} \right); \tag{3.16}$$

$$h_{\tilde{A}}^{\lambda} = \left(\bigcup_{[\gamma_{\tilde{A}}^{L}, \gamma_{\tilde{A}}^{U}] \in u_{\tilde{A}}} \left\{ [(\gamma_{\tilde{A}}^{L})^{\lambda}, (\gamma_{\tilde{A}}^{U})^{\lambda}] \right\}, \bigcup_{[\delta_{\tilde{A}}^{L}, \delta_{\tilde{A}}^{U}] \in w_{\tilde{A}}} \left\{ [1-(1-\delta_{\tilde{A}}^{L})^{\lambda}, 1-(1-\delta_{\tilde{A}}^{U})^{\lambda}] \right\}, \bigcup_{[\eta_{\tilde{A}}^{L}, \eta_{\tilde{A}}^{U}] \in v_{\tilde{A}}} \left\{ [1-(1-\eta_{\tilde{A}}^{L})^{\lambda}, 1-(1-\eta_{\tilde{A}}^{U})^{\lambda}] \right\} \right); \quad (3.17)$$

where $\lambda > 0$.

References

1. M. Akram, S. Naz, F. Smarandache, Generalization of maximizing deviation and TOPSIS method for MADM in simplified neutrosophic hesitant fuzzy environment. Symmetry **11**, 1058 (2019)
2. P. Juan-juan, W. Jian-qiang, H. Jun-hua, Multi-criteria decision making approach based on single-valued neutrosophic hesitant fuzzy geometric weighted Choquet integral Heronian mean operator. J. Intell. Fuzzy Syst. **2018**, 1–14 (2018)
3. P. Kakati, S. Borkotokey, S. Rahman, B. Davvaz, Interval neutrosophic hesitant fuzzy Einstein Choquet integral operator for multicriteria decision making. Artif. Intell. Rev. **53**, 2171–2206 (2020)
4. X. Li, X. Zhang, Single-valued neutrosophic hesitant fuzzy Choquet aggregation operators for multi-attribute decision making. Symmetry **10**, 50 (2018)
5. P. Liu, L. Shi, The generalized hybrid weighted average operator based on interval neutrosophic hesitant set and its application to multiple attribute decision making. Neural Comput. Appl. **26**, 457–471 (2015)
6. C.F. Liu, Y.S. Luo, New aggregation operators of single-valued neutrosophic hesitant fuzzy set and their application in multi-attribute decision making. Pattern Anal. Appl. **22**, 417–427 (2019)
7. F. Smarandache, *A Unifying Field in Logics Neutrosophic Logic. Neutrosophy, Neutrosophic Set, Neutrosophic Probability* (American Research Press, Champaign, 2003)
8. R. Sahin, P. Liu, Correlation coefficient of single-valued neutrosophic hesitant fuzzy sets and its applications in decision making. Neural Comput. Appl. **28**, 1387–1395 (2017)
9. S. Shao, X. Zhang, Y. Li, C. Bo, Probabilistic single-valued (interval) neutrosophic hesitant fuzzy set and its application in multi-attribute decision making. Symmetry **10**, 419 (2018)
10. R. Wang, Y. Li, Generalized single-valued neutrosophic hesitant fuzzy prioritized aggregation operators and their applications to multiple criteria decision making. Information **9**, 10 (2018)
11. J. Ye, Hesitant interval neutrosophic linguistic set and its application in multiple attribute decision making. Int. J. Mach. Learn. Cybern. **10**, 667–678 (2019)
12. J. Ye, Multiple-attribute decision making method under a single valued neutrosophic hesitant fuzzy environment. J. Intell. Syst. **24**, 23–36 (2015)
13. J. Ye, Correlation coefficients of interval neutrosophic hesitant fuzzy sets and its application in a multiple attribute decision making method. Informatica **27**, 179–202 (2016)

Chapter 4
Pythagorean Hesitant Fuzzy Set

Abstract Pythagorean hesitant fuzzy set is the first concept which sets up the proper framework for this chapter. Then, in order to put forward the issues and concepts described in this chapter, we give an overall view of interval-valued form of Pythagorean hesitant fuzzy set relevant to this contribution. The last concept which is needed for a complete characterization of Pythagorean hesitant fuzzy set is the dual Pythagorean hesitant fuzzy set.

4.1 Pythagorean Hesitant Fuzzy Set

By generalizing the concept of intuitionistic hesitant fuzzy set, Khan et al. [2] represented the concept of Pythagorean hesitant fuzzy set, and subsequently, they developed Pythagorean hesitant fuzzy weighted averaging operator and Pythagorean hesitant fuzzy weighted geometric operator for multiple criteria decision making problem. Further, Khan et al. proposed the ordered types of aforementioned operators in [3]. Pythagorean hesitant fuzzy set has been widely used in other fields of decision making, such as the study in this issue by Liang and Xu [4] which concerns presenting distance measures for Pythagorean hesitant fuzzy sets, and the study of Zhong et al. [8] in which the concentration is specified on the five parameters: membership degree, non-membership degree, indeterminacy degree, strength of commitment about membership, and direction of commitment.

By considering X as the reference set, Yager [6] defined a Pythagorean fuzzy set (PFS) A on X in terms of two functions u_A and v_A as follows:

$$A = \{\langle x, u_A(x), v_A(x)\rangle \mid x \in X\}, \tag{4.1}$$

where $u_A(x)$ and $v_A(x)$ are the sets of some different values in [0, 1] and represent membership and non-membership degrees of the element $x \in X$ to A, respectively.

B. Farhadinia, *Hesitant Fuzzy Set*, Computational Intelligence Methods and Applications, https://doi.org/10.1007/978-981-16-7301-6_4

Both membership and non-membership degrees satisfy

$$0 \le u_A^2(x) + v_A^2(x) \le 1, \tag{4.2}$$

and help to characterize the degree of indeterminacy $\pi_A(x) = \sqrt[2]{1 - u_A^2(x) + v_A^2(x)}$ for any $x \in X$.

Given a fixed $x \in X$, some operational laws on PFSs are defined as follows [6]:

$$A^c = \{\langle x, v_A(x), u_A(x)\rangle \mid x \in X\};$$

$$A_1 \cup A_2 = \{\langle x, \max\{u_{A_1}(x), u_{A_2}(x)\}, \min\{v_{A_1}(x), v_{A_2}(x)\}\rangle \mid x \in X\};$$

$$A_1 \cap A_2 = \{\langle x, \min\{u_{A_1}(x), u_{A_2}(x)\}, \max\{v_{A_1}(x), v_{A_2}(x)\}\rangle \mid x \in X\};$$

$$A_1 \oplus A_2 = \left\{\left\langle x, \sqrt[2]{u_{A_1}^2(x) + u_{A_2}^2(x) - u_{A_1}^2(x) \times u_{A_2}^2(x)}, v_{A_1}(x) \times v_{A_2}(x)\right\rangle \mid x \in X\right\};$$

$$A_1 \otimes A_2 = \left\{\left\langle x, u_{A_1}(x) \times u_{A_2}(x), \sqrt[2]{v_{A_1}^2(x) + v_{A_2}^2(x) - v_{A_1}^2(x) \times v_{A_2}^2(x)}\right\rangle \mid x \in X\right\};$$

$$\lambda A = \left\{\left\langle x, \sqrt[2]{1 - (1 - u_A^2(x))^\lambda}, (v_A(x))^\lambda\right\rangle \mid x \in X\right\}, \quad \lambda > 0;$$

$$A^\lambda = \left\{\left\langle x, (u_A(x))^\lambda, \sqrt[2]{1 - (1 - v_A^2(x))^\lambda},\right\rangle \mid x \in X\right\}, \quad \lambda > 0.$$

Example 4.1 Let $X = \{x_1, x_2\}$ be the reference set, $h_A(x_1) = \{(0.2, 0, 3)\}$ and $h_A(x_2) = \{(0.4, 0, 3)\}$. Then A can be considered as an PFS, i.e.,

$$A = \{\langle x_1, (0.2, 0, 3)\rangle, \langle x_2, (0.4, 0, 3)\rangle\}.$$

Definition 4.1 ([2]) Let X be a reference set. A Pythagorean hesitant fuzzy set (PHFS) A on X is defined in terms of two functions $u_A(x)$ and $v_A(x)$ as follows:

$$A = \{\langle x, u_A(x), v_A(x)\rangle \mid x \in X\}, \tag{4.3}$$

where $u_A(x)$ and $v_A(x)$ are the sets of some different values in [0, 1] and represent the possible membership degrees and non-membership degrees of the element $x \in X$ to A, respectively.

Here, for all $x \in X$, if we consider $u_A(x) = \bigcup_{\gamma_A \in u_A(x)}\{\gamma_A\}$, $v_A(x) = \bigcup_{\eta_A \in v_A(x)}\{\eta_A\}$, ${\gamma_A}^+ \in u_A^+ = \bigcup_{x \in X} \max_{\gamma_A \in u_A(x)}\{\gamma_A\}$ and ${\eta_A}^+ \in v_A^+ = \bigcup_{x \in X} \max_{\eta_A \in v_A(x)}\{\eta_A\}$, then we find that

$$0 \le \gamma_A, \ \eta_A \le 1, \quad 0 \le ({\gamma_A}^+)^2 + ({\eta_A}^+)^2 \le 1.$$

For the sake of simplicity, the pair

$$h_A(x) := (u_A(x), v_A(x)) = \left(\bigcup_{\gamma_A \in u_A(x)} \{\gamma_A\}, \bigcup_{\eta_A \in v_A(x)} \{\eta_A\} \right) \tag{4.4}$$

is called as the Pythagorean hesitant fuzzy element (PHFE).

Remark 4.1 Throughout this book, the set of all PHFSs on the reference set X is denoted by $\mathbb{PHFS}(X)$.

Before giving the definition of algebraic operations for PHFSs, let us discuss more or less about the complement operator for PHFSs. The complement of a PHFS A, denoted by A^c, is defined in form of (see [6])

$$\begin{aligned} A^c &= \{\langle x, u_A^c(x), v_A^c(x)\rangle \mid x \in X\} \\ &= \begin{cases} \{\langle x, \bigcup_{\eta_A \in v_A(x)}\{\eta_A\}, \bigcup_{\gamma_A \in u_A(x)}\{\gamma_A\}\rangle | x \in X\}, & \text{if } \ u_A \neq \emptyset, v_A \neq \emptyset; \\ \{\langle x, \bigcup_{\gamma_A \in u_A(x)}\{1-\gamma_A\}, \{\emptyset\}\rangle | x \in X\}, & \text{if } \ u_A \neq \emptyset, v_A = \emptyset; \\ \{\langle x, \{\emptyset\}, \bigcup_{\eta_A \in v_A(x)}\{1-\eta_A\}\rangle | x \in X\}, & \text{if } \ u_A = \emptyset, v_A \neq \emptyset \end{cases} \end{aligned}$$

such that

$$0 \leq {\gamma_A}^c, \ {\eta_A}^c \leq 1, \quad 0 \leq ({\gamma_A}^{c+})^2 + ({\eta_A}^{c+})^2 \leq 1.$$

Example 4.2 Let $X = \{x_1, x_2\}$ be the reference set, $h_A(x_1) = (u_A(x_1), v_A(x_1)) = (\{0.2, 0.5\}, \{0.3\})$ and $h_A(x_2) = (u_A(x_2), v_A(x_2)) = (\{0.3, 0.4\}, \{0.1, 0.6\})$. Then, $h_A(x_i)$ for $i = 1, 2$ are the PHFEs of x_i $(i = 1, 2)$ in the set A, because the relations

$$\begin{aligned} ({\gamma_A}^+)(x_1) = 0.5, \ ({\eta_A}^+)(x_1) = 0.3; \\ ({\gamma_A}^+)(x_2) = 0.4, \ ({\eta_A}^+)(x_2) = 0.6; \end{aligned}$$

result in

$$\begin{aligned} ({\gamma_A}^+)^2(x_1) + ({\eta_A}^+)^2(x_1) = 0.5^2 + 0.3^2 \leq 1; \\ ({\gamma_A}^+)^2(x_2) + ({\eta_A}^+)^2(x_2) = 0.4^2 + 0.6^2 \leq 1. \end{aligned}$$

Thus, A is a PHFS, and it is denoted by

$$A = \{\langle x_1, \{0.2, 0.5\}, \{0.3\}\rangle, \ \langle x_2, \{0.3, 0.4\}, \{0.1, 0.6\}\rangle\}.$$

Remark that for a given PHFE $h_A \neq \emptyset$, if u_A and v_A possess only one value γ_A and η_A, respectively, such that $0 \leq {\gamma_A}^2 + {\eta_A}^2 \leq 1$, then the PHFS reduces to an PFS [54]. If $u_A \neq \emptyset$ and $v_A = \emptyset$, then the PHFS reduces to a PFS.

For the PHFEs $h_A = (u_A, v_A)$, $h_{A_1} = (u_{A_1}, v_{A_1})$ and $h_{A_2} = (u_{A_2}, v_{A_2})$ the following operations are defined:

$$(h_A)^c = (v_A, u_A); \tag{4.5}$$

$$h_{A_1} \cup h_{A_2} = (u_{A_1} \cup u_{A_2}, v_{A_1} \cap v_{A_2}); \tag{4.6}$$

$$h_{A_1} \cap h_{A_2} = (u_{A_1} \cap u_{A_2}, v_{A_1} \cup v_{A_2}); \tag{4.7}$$

$$h_{A_1} \oplus h_{A_2} = (u_{A_1} \oplus u_{A_2}, v_{A_1} \otimes v_{A_2}); \tag{4.8}$$

$$h_{A_1} \otimes h_{A_2} = (u_{A_1} \otimes u_{A_2}, v_{A_1} \oplus v_{A_2}), \tag{4.9}$$

where

$$0 \le (\gamma_A{}^+)^2 + (\eta_A{}^+)^2 \le 1,\ 0 \le (\gamma_{A_1^+})^2 + (\eta_{A_1^+})^2 \le 1,\ 0 \le (\gamma_{A_2^+})^2 + (\eta_{A_2^+})^2 \le 1.$$

On the basis of the operations defined on PHFEs, some set and algebraic operations on PHFSs can be established as follows:

$$A_1 \cup A_2 = \bigcup_{h_{A_1} \in A_1, h_{A_2} \in A_2} h_{A_1} \cup h_{A_2}$$
$$= \{\langle x, u_{A_1}(x) \cup u_{A_2}(x), v_{A_1}(x) \cap v_{A_2}(x)\rangle | x \in X\}, \tag{4.10}$$

$$A_1 \cap A_2 = \bigcup_{h_{A_1} \in A_1, h_{A_2} \in A_2} h_{A_1} \cap h_{A_2}$$
$$= \{\langle x, u_{A_1}(x) \cap u_{A_2}(x), v_{A_1}(x) \cup v_{A_2}(x)\rangle | x \in X\}, \tag{4.11}$$

$$A_1 \oplus A_2 = \bigcup_{h_{A_1} \in A_1, h_{A_2} \in A_2} h_{A_1} \oplus h_{A_2}$$
$$= \{\langle x, u_{A_1}(x) \oplus u_{A_2}(x), v_{A_1}(x) \otimes v_{A_2}(x)\rangle | x \in X\}, \tag{4.12}$$

$$A_1 \otimes A_2 = \bigcup_{h_{A_1} \in A_1, h_{A_2} \in A_2} h_{A_1} \otimes h_{A_2}$$
$$= \{\langle x, u_{A_1}(x) \otimes u_{A_2}(x), v_{A_1}(x) \oplus v_{A_2}(x)\rangle | x \in X\}. \tag{4.13}$$

4.2 Interval-Valued Pythagorean Hesitant Fuzzy Set

The concept of interval-valued Pythagorean hesitant fuzzy set permits the membership and non-membership degrees of an element to a given set to

(continued)

have a few different interval values. By having this motivation, Wang et al. [5] expanded the concept of Pythagorean hesitant fuzzy set to interval-valued Pythagorean hesitant fuzzy set, and further developed a number of corresponding operational laws, score function, and generalized distance measures. Then, Zhang et al. [7] developed a series of interval-valued Pythagorean hesitant fuzzy aggregation operators for applying to multiple criteria group decision making problems.

Definition 4.2 ([7]) Let X be a reference set. An interval-valued Pythagorean hesitant fuzzy set (IVPHFS) $\widetilde{A}$ on X is defined in terms of two functions $u_{\widetilde{A}}(x)$ and $v_{\widetilde{A}}(x)$ as follows:

$$\widetilde{A} = \{\langle x, u_{\widetilde{A}}(x), v_{\widetilde{A}}(x)\rangle \mid x \in X\}, \tag{4.14}$$

where $u_{\widetilde{A}}(x)$ and $v_{\widetilde{A}}(x)$ are some different interval values in $[0, 1]$ and represent the possible interval membership degrees and interval non-membership degrees of the element $x \in X$ to the set $\widetilde{A}$, respectively.

Here, for all $x \in X$, if we take $u_{\widetilde{A}}(x) = \bigcup_{[\gamma_{\widetilde{A}}^L, \gamma_{\widetilde{A}}^U] \in u_{\widetilde{A}}(x)} \{[\gamma_{\widetilde{A}}^L, \gamma_{\widetilde{A}}^U]\}$, $v_{\widetilde{A}}(x) = \bigcup_{[\eta_{\widetilde{A}}^L, \eta_{\widetilde{A}}^U] \in v_{\widetilde{A}}(x)} \{[\eta_{\widetilde{A}}^L, \eta_{\widetilde{A}}^U]\}$, $\gamma_{\widetilde{A}}^{U+} \in \bigcup_{x \in X} \max\{\gamma_{\widetilde{A}}^U(x)\}$ and $\eta_{\widetilde{A}}^{U+} \in \bigcup_{x \in X} \max\{\eta_{\widetilde{A}}^U(x)\}$, then we conclude that

$$0 \le \gamma_{\widetilde{A}}^L, \gamma_{\widetilde{A}}^U, \ \eta_{\widetilde{A}}^L, \eta_{\widetilde{A}}^U \le 1, \quad 0 \le (\gamma_{\widetilde{A}}^{U+})^2 + (\eta_{\widetilde{A}}^{U+})^2 \le 1.$$

For the sake of simplicity, the pair

$$h_{\widetilde{A}}(x) := (u_{\widetilde{A}}(x), v_{\widetilde{A}}(x)) = \left(\bigcup_{[\gamma_{\widetilde{A}}^L, \gamma_{\widetilde{A}}^U] \in u_{\widetilde{A}}(x)} \{[\gamma_{\widetilde{A}}^L, \gamma_{\widetilde{A}}^U]\}, \bigcup_{[\eta_{\widetilde{A}}^L, \eta_{\widetilde{A}}^U] \in v_{\widetilde{A}}(x)} \{[\eta_{\widetilde{A}}^L, \eta_{\widetilde{A}}^U]\} \right) \tag{4.15}$$

is called as the interval-valued Pythagorean hesitant fuzzy element (IVPHFE).

Remark 4.2 Throughout this book, the set of all IVPHFSs on the reference set X is denoted by $\mathbb{IVPHFS}(X)$.

Example 4.3 Let $X = \{x_1, x_2\}$ be the reference set,
$h_{\widetilde{A}}(x_1) = (\bigcup_{[\gamma_{\widetilde{A}}^L, \gamma_{\widetilde{A}}^U] \in u_{\widetilde{A}}(x_1)} \{[\gamma_{\widetilde{A}}^L, \gamma_{\widetilde{A}}^U]\}, \bigcup_{[\eta_{\widetilde{A}}^L, \eta_{\widetilde{A}}^U] \in v_{\widetilde{A}}(x_1)} \{[\eta_{\widetilde{A}}^L, \eta_{\widetilde{A}}^U]\}) =$
$(\{[0.2, 0.5], [0.3, 0.3]\}, \{[0.3, 0.6]\})$ and
$h_{\widetilde{A}}(x_2) = (\bigcup_{[\gamma_{\widetilde{A}}^L, \gamma_{\widetilde{A}}^U] \in u_{\widetilde{A}}(x_2)} \{[\gamma_{\widetilde{A}}^L, \gamma_{\widetilde{A}}^U]\}, \bigcup_{[\eta_{\widetilde{A}}^L, \eta_{\widetilde{A}}^U] \in v_{\widetilde{A}}(x_2)} \{[\eta_{\widetilde{A}}^L, \eta_{\widetilde{A}}^U]\}) =$

$(\{[0.3, 0.4], [0.1, 0.6]\}, \{[0.2, 0.5]\})$ be the IVPHFEs of x_i $(i = 1, 2)$ in the set $\widetilde{A}$, respectively. In this case,

$$(\gamma_{\widetilde{A}}^{L+})(x_1) = 0.3,\ (\eta_{\widetilde{A}}^{U+})(x_1) = 0.6;$$
$$(\gamma_{\widetilde{A}}^{L+})(x_2) = 0.3,\ (\eta_{\widetilde{A}}^{U+})(x_2) = 0.6;$$

result in

$$(\gamma_{\widetilde{A}}^{L+})^2(x_1) + (\eta_{\widetilde{A}}^{U+})^2(x_1) = 0.3^2 + 0.6^2 \leq 1;$$
$$(\gamma_{\widetilde{A}}^{L+})^2(x_2) + (\eta_{\widetilde{A}}^{U+})^2(x_2) = 0.3^2 + 0.6^2 \leq 1.$$

Thus, $h_{\widetilde{A}}(x_1)$ and $h_{\widetilde{A}}(x_2)$ are two IVPHFEs, and then $\widetilde{A}$ can be considered as a IVPHFS, i.e.,

$$\widetilde{A} = \{\langle x_1, \{[0.2, 0.5], [0.3, 0.3]\}, \{[0.3, 0.6]\}\rangle,\ \langle x_2, \{[0.3, 0.4], [0.1, 0.6]\}, \{[0.2, 0.5]\}\rangle\}.$$

For the IVPHFEs $h_{\widetilde{A}} = (u_{\widetilde{A}}, v_{\widetilde{A}})$, $h_{\widetilde{A}_1} = (u_{\widetilde{A}_1}, v_{\widetilde{A}_1})$ and $h_{\widetilde{A}_2} = (u_{\widetilde{A}_2}, v_{\widetilde{A}_2})$ the following operations are defined (see [7]):

$$h_{\widetilde{A}_1} \oplus h_{\widetilde{A}_2} = \left(\bigcup_{[\gamma_{\widetilde{A}_1}^L, \gamma_{\widetilde{A}_1}^U] \in u_{\widetilde{A}_1}} \left\{\left[\gamma_{\widetilde{A}_1}^L, \gamma_{\widetilde{A}_1}^U\right]\right\}, \bigcup_{[\eta_{\widetilde{A}_1}^L, \eta_{\widetilde{A}_1}^U] \in v_{\widetilde{A}_1}} \left\{\left[\eta_{\widetilde{A}_1}^L, \eta_{\widetilde{A}_1}^U\right]\right\} \right)$$
$$\oplus \left(\bigcup_{[\gamma_{\widetilde{A}_2}^L, \gamma_{\widetilde{A}_2}^U] \in u_{\widetilde{A}_2}} \left\{\left[\gamma_{\widetilde{A}_2}^L, \gamma_{\widetilde{A}_2}^U\right]\right\}, \bigcup_{[\eta_{\widetilde{A}_2}^L, \eta_{\widetilde{A}_2}^U] \in v_{\widetilde{A}_2}} \left\{\left[\eta_{\widetilde{A}_2}^L, \eta_{\widetilde{A}_2}^U\right]\right\} \right)$$
$$= \left(\bigcup_{[\gamma_{\widetilde{A}_1}^L, \gamma_{\widetilde{A}_1}^U] \in u_{\widetilde{A}_1}, [\gamma_{\widetilde{A}_2}^L, \gamma_{\widetilde{A}_2}^U] \in u_{\widetilde{A}_2}} \left\{\left[\sqrt[2]{(\gamma_{\widetilde{A}_1}^L)^2 + (\gamma_{\widetilde{A}_2}^L)^2 - (\gamma_{\widetilde{A}_1}^L)^2 \times (\gamma_{\widetilde{A}_2}^L)^2}, \right.\right.\right.$$
$$\left.\left.\sqrt[2]{(\gamma_{\widetilde{A}_1}^U)^2 + (\gamma_{\widetilde{A}_2}^U)^2 - (\gamma_{\widetilde{A}_1}^U)^2 \times (\gamma_{\widetilde{A}_2}^U)^2}\right]\right\},$$
$$\left. \bigcup_{[\eta_{\widetilde{A}_1}^L, \eta_{\widetilde{A}_1}^U] \in v_{\widetilde{A}_1}, [\eta_{\widetilde{A}_2}^L, \eta_{\widetilde{A}_2}^U] \in v_{\widetilde{A}_2}} \left\{\left[\eta_{\widetilde{A}_1}^L \times \eta_{\widetilde{A}_2}^L, \eta_{\widetilde{A}_1}^U \times \eta_{\widetilde{A}_2}^U\right]\right\} \right); \tag{4.16}$$

$$h_{\widetilde{A}_1} \otimes h_{\widetilde{A}_2} = \left(\bigcup_{[\gamma^L_{\widetilde{A}_1}, \gamma^U_{\widetilde{A}_1}] \in u_{\widetilde{A}_1}} \left\{ \left[\gamma^L_{\widetilde{A}_1}, \gamma^U_{\widetilde{A}_1} \right] \right\}, \bigcup_{[\eta^L_{\widetilde{A}_1}, \eta^U_{\widetilde{A}_1}] \in v_{\widetilde{A}_1}} \left\{ \left[\eta^L_{\widetilde{A}_1}, \eta^U_{\widetilde{A}_1} \right] \right\} \right)$$

$$\otimes \left(\bigcup_{[\gamma^L_{\widetilde{A}_2}, \gamma^U_{\widetilde{A}_2}] \in u_{\widetilde{A}_2}} \left\{ \left[\gamma^L_{\widetilde{A}_2}, \gamma^U_{\widetilde{A}_2} \right] \right\}, \bigcup_{[\eta^L_{\widetilde{A}_2}, \eta^U_{\widetilde{A}_2}] \in v_{\widetilde{A}_2}} \left\{ \left[\eta^L_{\widetilde{A}_2}, \eta^U_{\widetilde{A}_2} \right] \right\} \right)$$

$$= \left(\bigcup_{[\gamma^L_{\widetilde{A}_1}, \gamma^U_{\widetilde{A}_1}] \in v_{\widetilde{A}_1}, [\gamma^L_{\widetilde{A}_2}, \gamma^U_{\widetilde{A}_2}] \in v_{\widetilde{A}_2}} \left\{ \left[\gamma^L_{\widetilde{A}_1} \times \gamma^L_{\widetilde{A}_2}, \gamma^U_{\widetilde{A}_1} \times \gamma^U_{\widetilde{A}_2} \right] \right\}, \right.$$

$$\bigcup_{[\eta^L_{\widetilde{A}_1}, \eta^U_{\widetilde{A}_1}] \in u_{\widetilde{A}_1}, [\eta^L_{\widetilde{A}_2}, \eta^U_{\widetilde{A}_2}] \in u_{\widetilde{A}_2}} \left\{ \left[\sqrt[2]{(\eta^L_{\widetilde{A}_1})^2 + (\eta^L_{\widetilde{A}_2})^2 - (\eta^L_{\widetilde{A}_1})^2 \times (\eta^L_{\widetilde{A}_2})^2}, \right. \right.$$

$$\left. \left. \left. \sqrt[2]{(\eta^U_{\widetilde{A}_1})^2 + (\eta^U_{\widetilde{A}_2})^2 - (\eta^U_{\widetilde{A}_1})^2 \times (\eta^U_{\widetilde{A}_2})^2} \right] \right\} \right); \tag{4.17}$$

$$\lambda h_{\widetilde{A}} = \left(\bigcup_{[\gamma^L_{\widetilde{A}}, \gamma^U_{\widetilde{A}}] \in u_{\widetilde{A}}} \left\{ \left[\sqrt[2]{1 - (1 - (\gamma^L_{\widetilde{A}})^2)^\lambda}, \sqrt[2]{1 - (1 - (\gamma^U_{\widetilde{A}})^2)^\lambda} \right] \right\}, \right.$$

$$\left. \bigcup_{[\eta^L_{\widetilde{A}}, \eta^U_{\widetilde{A}}] \in v_{\widetilde{A}}} \left\{ \left[(\eta^L_{\widetilde{A}})^\lambda, (\eta^U_{\widetilde{A}})^\lambda \right] \right\} \right); \tag{4.18}$$

$$h_{\widetilde{A}}^\lambda = \left(\bigcup_{[\gamma^L_{\widetilde{A}}, \gamma^U_{\widetilde{A}}] \in u_{\widetilde{A}}} \left\{ \left[(\gamma^L_{\widetilde{A}})^\lambda, (\gamma^U_{\widetilde{A}})^\lambda \right] \right\}, \right.$$

$$\left. \bigcup_{[\eta^L_{\widetilde{A}}, \eta^U_{\widetilde{A}}] \in v_{\widetilde{A}}} \left\{ \left[\sqrt[2]{1 - (1 - (\eta^L_{\widetilde{A}})^2)^\lambda}, \sqrt[2]{1 - (1 - (\eta^U_{\widetilde{A}})^2)^\lambda} \right] \right\} \right). \tag{4.19}$$

4.3 Dual Pythagorean Hesitant Fuzzy Set

Remark 4.3 Although, some contributions (for instance, [1]) have introduced the concept of dual Pythagorean hesitant fuzzy set, they indeed have dealt with the re-phrased form of Pythagorean hesitant fuzzy set which has already been discussed in the previous section. Therefore, nothing is added here concerning that concept.

References

1. X. Ji, L. Yu, J. Fu, Evaluating personal default risk in P2P lending platform: based on dual hesitant Pythagorean fuzzy TODIM approach. Mathematics **8**, 8 (2020)
2. M.S.A. Khan, S. Abdullah, A. Ali, N. Siddiqui, F. Amin, Pythagorean hesitant fuzzy sets and their application to group decision making with incomplete weight information. J. Intell. Fuzzy Syst. **33**, 3971–3985 (2017)
3. M.S.A. Khan, M.Y. Ali, S. Abdullah, I. Hussain, M. Farooq, Extension of TOPSIS method base on Choquet integral under interval-valued Pythagorean fuzzy environment. J. Intell. Fuzzy Syst. **34**, 267–282 (2018)
4. D. Liang, Z. Xu, The new extension of TOPSIS method for multiple criteria decision making with hesitant Pythagorean fuzzy sets. Appl. Soft Comput. **60**, 167–179 (2017)
5. L. Wang, H. Wang, Z. Xu, Z. Ren, The interval-valued hesitant pythagorean fuzzy set and its applications with extended TOPSIS and Choquet integral-based method. Int. J. Intell. Syst. **34**(6), 1063–1085 (2019)
6. R.R. Yager, Pythagorean fuzzy subsets, in *Proc Joint IFSA World Congress and NAFIPS*, p. 9
7. M.Y. Zhang, T.T. Zheng, W.R. Zheng, L.G. Zhou, Interval-valued Pythagorean hesitant fuzzy set and its application to multi-attribute group decision making. Complexity (2020). https://doi.org/10.1155/2020/1724943
8. Y. Zhong, X. Guo, H. Gao, M. Huang, A new distance measure based on Pythagorean hesitant fuzzy sets and its application to multi-criteria decision making, in *Proc. SPIE 11321, 2019 International Conference on Image and Video Processing, and Artificial Intelligence*. https://doi.org/10.1117/12.2541318

Chapter 5
q-Rung Orthopair Hesitant Fuzzy Set

Abstract q-rung orthopair hesitant fuzzy set is the first concept which sets up the proper framework for this chapter. Then, in order to put forward the issues and concepts described in this chapter, we give an overall view of interval-valued form of q-rung orthopair hesitant fuzzy set relevant to this contribution. The last concept which is needed for a complete characterization of q-rung orthopair hesitant fuzzy set is the dual q-rung orthopair hesitant fuzzy set.

5.1 q-Rung Orthopair Hesitant Fuzzy Set

Liu et al. [3] proposed a number of operations, score, and accurate functions for q-rung orthopair hesitant fuzzy sets together with a ranking technique of q-rung orthopair hesitant fuzzy sets. Furthermore, they established a distance measure for q-rung orthopair hesitant fuzzy sets in order to handle the uncertainty involved in a related TOPSIS approach. Moreover, Wang et al. [6] developed some distance and similarity measures of q-rung orthopair hesitant fuzzy sets and investigated their properties. In addition to that, they defined the axiomatic form of entropy measure for q-rung orthopair hesitant fuzzy sets. Hussain et al. [2] developed a series of operations, score, and accurate functions together with comparison rule for q-rung orthopair hesitant fuzzy sets, and then implemented them in a q-rung orthopair hesitant fuzzy-based decision making process.

By considering the reference set X, Yager [8] proposed a q-rung orthopair fuzzy set (q-ROFS) A on X in terms of two functions u_A and v_A as follows:

$$A = \{\langle x, u_A(x), v_A(x)\rangle \mid x \in X\}, \tag{5.1}$$

B. Farhadinia, *Hesitant Fuzzy Set*, Computational Intelligence Methods and Applications, https://doi.org/10.1007/978-981-16-7301-6_5

where $u_A(x)$ and $v_A(x)$ are the sets of some different values in [0, 1] and represent membership and non-membership degrees of the element $x \in X$ to A, respectively. Both membership and non-membership degrees satisfy

$$0 \leq u_A^q(x) + v_A^q(x) \leq 1, \tag{5.2}$$

where $q > 1$, and help to characterize the degree of indeterminacy $\pi_A(x) = \sqrt[q]{1 - u_A^q(x) + v_A^q(x)}$ for any $x \in X$.

It is worth mentioning that any q-ROFS is the generalization of intuitionistic fuzzy set and Pythagorean fuzzy set. Whenever $q = 1$, the q-ROFS degenerates to an intuitionistic fuzzy set, and whenever $q = 2$, the q-ROFS degenerates to a Pythagorean fuzzy set.

Given a fixed $x \in X$, some operational laws on q-ROFSs are defined as follows (see [4]):

$$\begin{aligned}
A^c &= \{\langle x, v_A(x), u_A(x)\rangle \mid x \in X\};\\
A_1 \cup A_2 &= \{\langle x, \max\{u_{A_1}(x), u_{A_2}(x)\}, \min\{v_{A_1}(x), v_{A_2}(x)\}\rangle \mid x \in X\};\\
A_1 \cap A_2 &= \left\{\langle x, \min\{u_{A_1}(x), u_{A_2}(x)\}, \max\{v_{A_1}(x), v_{A_2}(x)\}\rangle \mid x \in X\right\};\\
A_1 \oplus A_2 &= \left\{\langle x, \sqrt[q]{u_{A_1}^q(x) + u_{A_2}^q(x) - u_{A_1}^q(x) \times u_{A_2}^q(x)}, v_{A_1}(x) \times v_{A_2}(x)\rangle \mid x \in X\right\};\\
A_1 \otimes A_2 &= \left\{\langle x, u_{A_1}(x) \times u_{A_2}(x), \sqrt[q]{v_{A_1}^q(x) + v_{A_2}^q(x) - v_{A_1}^q(x) \times v_{A_2}^q(x)}\rangle \mid x \in X\right\};\\
\lambda A &= \left\{\langle x, \sqrt[q]{1 - (1 - u_A^q(x))^\lambda}, (v_A(x))^\lambda\rangle \mid x \in X\right\}, \quad \lambda > 0;\\
A^\lambda &= \left\{\langle x, (u_A(x))^\lambda, \sqrt[q]{1 - (1 - v_A^q(x))^\lambda}, \rangle \mid x \in X\right\}, \quad \lambda > 0,
\end{aligned}$$

where $q > 1$.

Example 5.1 Let $X = \{x_1, x_2\}$ be the reference set, $h_A(x_1) = \{(0.2, 0.3)\}$ and $h_A(x_2) = \{(0.8, 0.7)\}$. Clearly, we observe that the element $(0.8, 0.7)$ cannot be described by using neither intuitionistic fuzzy set nor Pythagorean fuzzy set, but $0.8^3 + 0.7^3 \leq 1$. This verifies the set $\widetilde{A}$ that is given by

$$\widetilde{A} = \{\langle x_1, (0.2, 0.3)\rangle, \langle x_2, (0.8, 0.7)\rangle\}$$

is a q-ROFS for $q \geq 3$.

Definition 5.1 ([3]) Let X be a reference set. A q-rung orthopair hesitant fuzzy set (q-ROHFS) A on X is defined in terms of two functions $u_A(x)$ and $v_A(x)$ as follows:

$$A = \{\langle x, u_A(x), v_A(x)\rangle \mid x \in X\},$$

where $u_A(x)$ and $v_A(x)$ are the sets of some different values in [0, 1] and represent the possible membership degrees and non-membership degrees of the element $x \in X$ to A, respectively.

Here, for all $x \in X$, if we consider $u_A(x) = \bigcup_{\gamma_A \in u_A(x)}\{\gamma_A\}$, $v_A(x) = \bigcup_{\eta_A \in v_A(x)}\{\eta_A\}$, $\gamma_A{}^+ \in u_A^+ = \bigcup_{x\in X} \max_{\gamma_A \in u_A(x)}\{\gamma_A\}$ and $\eta_A{}^+ \in v_A^+ = \bigcup_{x\in X} \max_{\eta_A \in v_A(x)}\{\eta_A\}$, then we conclude that

$$0 \leq \gamma_A,\ \eta_A \leq 1, \quad 0 \leq (\gamma_A{}^+)^q + (\eta_A{}^+)^q \leq 1, \quad q \geq 1.$$

For the sake of simplicity, the pair

$$h_A(x) := (u_A(x), v_A(x)) = (\bigcup_{\gamma_A \in u_A(x)} \{\gamma_A\}, \bigcup_{\eta_A \in v_A(x)} \{\eta_A\}) \tag{5.3}$$

is called the q-rung orthopair hesitant fuzzy element (q-ROHFE).

Remark 5.1 Throughout this book, the set of all q-ROHFSs on the reference set X is denoted by $q - \mathbb{ROHFS}(X)$.

Before giving the definition of algebraic operations for q-ROHFSs, let us discuss more or less about the complement operator for q-ROHFSs. The complement of a q-ROHFS A, denoted by A^c, is defined in form of

$$\begin{aligned} A^c &= \{\langle x, u_A^c(x), v_A^c(x)\rangle \mid x \in X\} \\ &= \begin{cases} \{\langle x, \bigcup_{\eta_A \in v_A(x)}\{\eta_A\}, \bigcup_{\gamma_A \in u_A(x)}\{\gamma_A\}\rangle | x \in X\}, & \text{if} \quad u_A \neq \emptyset, v_A \neq \emptyset; \\ \{\langle x, \bigcup_{\gamma_A \in u_A(x)}\{1-\gamma_A\}, \{\emptyset\}\rangle | x \in X\}, & \text{if} \quad u_A \neq \emptyset, v_A = \emptyset; \\ \{\langle x, \{\emptyset\}, \bigcup_{\eta_A \in v_A(x)}\{1-\eta_A\}\rangle | x \in X\}, & \text{if} \quad u_A = \emptyset, v_A \neq \emptyset \end{cases} \end{aligned}$$

such that

$$0 \leq \gamma_A{}^c,\ \eta_A{}^c \leq 1, \quad 0 \leq (\gamma_A{}^{c+})^q + (\eta_A{}^{c+})^q \leq 1, \quad q \geq 1.$$

Example 5.2 Let $X = \{x_1, x_2\}$ be the reference set, $h_A(x_1) = (u_A(x_1), v_A(x_1)) = (\{0.2, 0.7\}, \{0.8\})$ and $h_A(x_2) = (u_A(x_2), v_A(x_2)) = (\{0.3, 0.4\}, \{0.1, 0.6\})$. Then, $h_A(x_i)$ for $i = 1, 2$ are the q-ROHFSs of x_i $(i = 1, 2)$ in the set A, because the relations

$$(\gamma_A{}^+)(x_1) = 0.7, \ (\eta_A{}^+)(x_1) = 0.8;$$
$$(\gamma_A{}^+)(x_2) = 0.4, \ (\eta_A{}^+)(x_2) = 0.6;$$

result in

$$(\gamma_A{}^+)^2(x_1) + (\eta_A{}^+)^2(x_1) = 0.7^3 + 0.8^3 \leq 1;$$
$$(\gamma_A{}^+)^2(x_2) + (\eta_A{}^+)^2(x_2) = 0.4^3 + 0.6^3 \leq 1.$$

Thus, A is a q-ROHFS (with at least $q \geq 3$), and it is denoted by

$$A = \{\langle x_1, \{0.2, 0.7\}, \{0.8\}\rangle, \ \langle x_2, \{0.3, 0.4\}, \{0.1, 0.6\}\rangle\}.$$

Remark that for a given q-ROHFE $h_A \neq \emptyset$, if u_A and v_A possess only one value γ_A and η_A, respectively, such that $0 \leq {\gamma_A}^q + {\eta_A}^q \leq 1$, then the q-ROHFS reduces to an q-ROHFS. If $u_A \neq \emptyset$ and $v_A = \emptyset$, then the q-ROHFS reduces to a q-ROFS.

For the q-ROHFEs $h_A = (u_A, v_A)$, $h_{A_1} = (u_{A_1}, v_{A_1})$, and $h_{A_2} = (u_{A_2}, v_{A_2})$ the following operations are defined:

$$(h_A)^c = (v_A, u_A); \tag{5.4}$$

$$h_{A_1} \cup h_{A_2} = (u_{A_1} \cup u_{A_2}, v_{A_1} \cap v_{A_2}); \tag{5.5}$$

$$h_{A_1} \cap h_{A_2} = (u_{A_1} \cap u_{A_2}, v_{A_1} \cup v_{A_2}); \tag{5.6}$$

$$h_{A_1} \oplus h_{A_2} = (u_{A_1} \oplus u_{A_2}, v_{A_1} \otimes v_{A_2}); \tag{5.7}$$

$$h_{A_1} \otimes h_{A_2} = (u_{A_1} \otimes u_{A_2}, v_{A_1} \oplus v_{A_2}), \tag{5.8}$$

where

$$0 \leq ({\gamma_A}^+)^q + ({\eta_A}^+)^q \leq 1, \quad 0 \leq (\gamma_{A_1^+})^q + (\eta_{A_1^+})^q \leq 1,$$
$$0 \leq (\gamma_{A_2^+})^q + (\eta_{A_2^+})^q \leq 1,$$

for all $q \geq 1$.

On the basis of the operations defined on q-ROHFEs, some relationships can be further established for such operations on q-ROHFSs as follows:

$$A_1 \cup A_2 = \bigcup_{h_{A_1} \in A_1, h_{A_2} \in A_2} h_{A_1} \cup h_{A_2}$$
$$= \{\langle x, u_{A_1}(x) \cup u_{A_2}(x), v_{A_1}(x) \cap v_{A_2}(x)\rangle | x \in X\}, \tag{5.9}$$

$$A_1 \cap A_2 = \bigcup_{h_{A_1} \in A_1, h_{A_2} \in A_2} h_{A_1} \cap h_{A_2}$$
$$= \{\langle x, u_{A_1}(x) \cap u_{A_2}(x), v_{A_1}(x) \cup v_{A_2}(x)\rangle | x \in X\}, \tag{5.10}$$

$$A_1 \oplus A_2 = \bigcup_{h_{A_1} \in A_1, h_{A_2} \in A_2} h_{A_1} \oplus h_{A_2}$$
$$= \{\langle x, u_{A_1}(x) \oplus u_{A_2}(x), v_{A_1}(x) \otimes v_{A_2}(x)\rangle | x \in X\}, \tag{5.11}$$

$$A_1 \otimes A_2 = \bigcup_{h_{A_1} \in A_1, h_{A_2} \in A_2} h_{A_1} \otimes h_{A_2}$$
$$= \{\langle x, u_{A_1}(x) \otimes u_{A_2}(x), v_{A_1}(x) \oplus v_{A_2}(x)\rangle | x \in X\}. \tag{5.12}$$

5.2 Interval-Valued q-Rung Orthopair Hesitant Fuzzy Set

In [7], Xu et al. presented some operations, comparison technique, and aggregation operators of interval-valued q-rung dual hesitant fuzzy sets. Then, Feng et al. [1] extended the operator of power Hamy mean to operators for interval-valued q-rung dual hesitant fuzzy sets, called the interval-valued q-rung dual hesitant fuzzy power Hamy mean and the interval-valued q-rung dual hesitant fuzzy power weighted Hamy mean. Xu et al. [7] presented a number of operations and a comparison technique for interval-valued q-rung dual hesitant fuzzy sets, and then they developed some interval-valued q-rung dual hesitant fuzzy aggregation operators.

Definition 5.2 ([1]) Let X be a reference set. An interval-valued q-rung orthopair hesitant fuzzy set (IVq-ROHFS) $\widetilde{A}$ on X is defined in terms of two functions $u_{\widetilde{A}}(x)$ and $v_{\widetilde{A}}(x)$ as follows:

$$\widetilde{A} = \{\langle x, u_{\widetilde{A}}(x), v_{\widetilde{A}}(x)\rangle \mid x \in X\}, \tag{5.13}$$

where $u_{\widetilde{A}}(x)$ and $v_{\widetilde{A}}(x)$ are some different interval values in $[0, 1]$ and represent the possible interval membership degrees and interval non-membership degrees of the element $x \in X$ to the set $\widetilde{A}$, respectively.

Here, for all $x \in X$, if we consider $u_{\widetilde{A}}(x) = \bigcup_{[\gamma^L_{\widetilde{A}}, \gamma^U_{\widetilde{A}}] \in u_{\widetilde{A}}(x)} \{[\gamma^L_{\widetilde{A}}, \gamma^U_{\widetilde{A}}]\}$, $v_{\widetilde{A}}(x) = \bigcup_{[\eta^L_{\widetilde{A}}, \eta^U_{\widetilde{A}}] \in v_{\widetilde{A}}(x)} \{[\eta^L_{\widetilde{A}}, \eta^U_{\widetilde{A}}]\}$, $\gamma^{U+}_{\widetilde{A}} \in \bigcup_{x \in X} \max\{\gamma^U_{\widetilde{A}}(x)\}$ and $\eta^{U+}_{\widetilde{A}} \in \bigcup_{x \in X} \max\{\eta^U_{\widetilde{A}}(x)\}$, then we find that

$$0 \le \gamma^L_{\widetilde{A}}, \gamma^U_{\widetilde{A}},\ \eta^L_{\widetilde{A}}, \eta^U_{\widetilde{A}} \le 1, \quad 0 \le \left(\gamma^{U+}_{\widetilde{A}}\right)^q + \left(\eta^{U+}_{\widetilde{A}}\right)^q \le 1, \quad q \ge 1.$$

For the sake of simplicity, the pair

$$h_{\widetilde{A}}(x) := (u_{\widetilde{A}}(x), v_{\widetilde{A}}(x)) = \left(\bigcup_{[\gamma^L_{\widetilde{A}}, \gamma^U_{\widetilde{A}}] \in u_{\widetilde{A}}(x)} \{[\gamma^L_{\widetilde{A}}, \gamma^U_{\widetilde{A}}]\}, \bigcup_{[\eta^L_{\widetilde{A}}, \eta^U_{\widetilde{A}}] \in v_{\widetilde{A}}(x)} \{[\eta^L_{\widetilde{A}}, \eta^U_{\widetilde{A}}]\} \right) \tag{5.14}$$

is called the interval-valued q-rung orthopair hesitant fuzzy element (IVq-ROHFE).

Remark 5.2 Throughout this book, the set of all IVq-ROHFSs on the reference set X is denoted by $\mathbb{IV}q - \mathbb{ROHFS}(X)$.

Example 5.3 Let $X = \{x_1, x_2\}$ be the reference set,
$h_{\widetilde{A}}(x_1) = (\bigcup_{[\gamma^L_{\widetilde{A}}, \gamma^U_{\widetilde{A}}] \in u_{\widetilde{A}}(x_1)} \{[\gamma^L_{\widetilde{A}}, \gamma^U_{\widetilde{A}}]\}, \bigcup_{[\eta^L_{\widetilde{A}}, \eta^U_{\widetilde{A}}] \in v_{\widetilde{A}}(x_1)} \{[\eta^L_{\widetilde{A}}, \eta^U_{\widetilde{A}}]\})$
$= (\{[0.2, 0.5], [0.3, 0.3]\}, \{[0.7, 0.8]\})$ and
$h_{\widetilde{A}}(x_2) = (\bigcup_{[\gamma^L_{\widetilde{A}}, \gamma^U_{\widetilde{A}}] \in u_{\widetilde{A}}(x_2)} \{[\gamma^L_{\widetilde{A}}, \gamma^U_{\widetilde{A}}]\}, \bigcup_{[\eta^L_{\widetilde{A}}, \eta^U_{\widetilde{A}}] \in v_{\widetilde{A}}(x_2)} \{[\eta^L_{\widetilde{A}}, \eta^U_{\widetilde{A}}]\})$
$= (\{[0.3, 0.4], [0.1, 0.6]\}, \{[0.2, 0.5]\})$ be the IVPHFEs of x_i $(i = 1, 2)$ in the set $\widetilde{A}$, respectively. In this case,

$$\left(\gamma^{L+}_{\widetilde{A}}\right)(x_1) = 0.7, \ \left(\eta^{U+}_{\widetilde{A}}\right)(x_1) = 0.8;$$
$$\left(\gamma^{L+}_{\widetilde{A}}\right)(x_2) = 0.3, \ \left(\eta^{U+}_{\widetilde{A}}\right)(x_2) = 0.6;$$

result in

$$\left(\gamma^{L+}_{\widetilde{A}}\right)^q(x_1) + \left(\eta^{U+}_{\widetilde{A}}\right)^q(x_1) = 0.7^q + 0.8^q \le 1;$$
$$\left(\gamma^{L+}_{\widetilde{A}}\right)^q(x_2) + \left(\eta^{U+}_{\widetilde{A}}\right)^q(x_2) = 0.3^q + 0.6^q \le 1,$$

which hold true for any $q \ge 3$. Thus, $h_{\widetilde{A}}(x_1)$ and $h_{\widetilde{A}}(x_2)$ are two IV3-ROHFEs, and then $\widetilde{A}$ can be considered as a IV3-ROHFS, i.e.,

$$\widetilde{A} = \{\langle x_1, \{[0.2, 0.5], [0.3, 0.3]\}, \{[0.7, 0.8]\}\rangle, \ \langle x_2, \{[0.3, 0.4], [0.1, 0.6]\}, \{[0.2, 0.5]\}\rangle\}.$$

For the IVq-ROHFEs
$h_{\widetilde{A}} = (u_{\widetilde{A}}, v_{\widetilde{A}}) = (\bigcup_{[\gamma^L_{\widetilde{A}}, \gamma^U_{\widetilde{A}}] \in u_{\widetilde{A}}} \{[\gamma^L_{\widetilde{A}}, \gamma^U_{\widetilde{A}}]\}, \bigcup_{[\eta^L_{\widetilde{A}}, \eta^U_{\widetilde{A}}] \in v_{\widetilde{A}}} \{[\eta^L_{\widetilde{A}}, \eta^U_{\widetilde{A}}]\})$,
$h_{\widetilde{A}_1} = (u_{\widetilde{A}_1}, v_{\widetilde{A}_1}) = (\bigcup_{[\gamma^L_{\widetilde{A}_1}, \gamma^U_{\widetilde{A}_1}] \in u_{\widetilde{A}_1}} \{[\gamma^L_{\widetilde{A}_1}, \gamma^U_{\widetilde{A}_1}]\}, \bigcup_{[\eta^L_{\widetilde{A}_1}, \eta^U_{\widetilde{A}_1}] \in v_{\widetilde{A}_1}} \{[\eta^L_{\widetilde{A}_1}, \eta^U_{\widetilde{A}_1}]\})$ and
$h_{\widetilde{A}_2} = (u_{\widetilde{A}_2}, v_{\widetilde{A}_2}) = (\bigcup_{[\gamma^L_{\widetilde{A}_2}, \gamma^U_{\widetilde{A}_2}] \in u_{\widetilde{A}_2}} \{[\gamma^L_{\widetilde{A}_2}, \gamma^U_{\widetilde{A}_2}]\}, \bigcup_{[\eta^L_{\widetilde{A}_2}, \eta^U_{\widetilde{A}_2}] \in v_{\widetilde{A}_2}} \{[\eta^L_{\widetilde{A}_2}, \eta^U_{\widetilde{A}_2}]\})$ the following operations are defined (see [1]):

$$h_{\widetilde{A}_1} \oplus h_{\widetilde{A}_2} = \left(\bigcup_{[\gamma^L_{\widetilde{A}_1}, \gamma^U_{\widetilde{A}_1}] \in u_{\widetilde{A}_1}, [\gamma^L_{\widetilde{A}_2}, \gamma^U_{\widetilde{A}_2}] \in u_{\widetilde{A}_2}} \{[\sqrt[q]{(\gamma^L_{\widetilde{A}_1})^q + (\gamma^L_{\widetilde{A}_2})^q - (\gamma^L_{\widetilde{A}_1})^q \times (\gamma^L_{\widetilde{A}_2})^q}, \right.$$
$$\sqrt[q]{(\gamma^U_{\widetilde{A}_1})^q + (\gamma^U_{\widetilde{A}_2})^q - (\gamma^U_{\widetilde{A}_1})^q \times (\gamma^U_{\widetilde{A}_2})^q}]\},$$
$$\left. \bigcup_{[\eta^L_{\widetilde{A}_1}, \eta^U_{\widetilde{A}_1}] \in v_{\widetilde{A}_1}, [\eta^L_{\widetilde{A}_2}, \eta^U_{\widetilde{A}_2}] \in v_{\widetilde{A}_2}} \{[\eta^L_{\widetilde{A}_1} \times \eta^L_{\widetilde{A}_2}, \eta^U_{\widetilde{A}_1} \times \eta^U_{\widetilde{A}_2}]\} \right); \tag{5.15}$$

$$h_{\widetilde{A}_1} \otimes h_{\widetilde{A}_2} = \left(\bigcup_{[\gamma^L_{\widetilde{A}_1},\gamma^U_{\widetilde{A}_1}]\in v_{\widetilde{A}_1},[\gamma^L_{\widetilde{A}_2},\gamma^U_{\widetilde{A}_2}]\in v_{\widetilde{A}_2}} \{[\gamma^L_{\widetilde{A}_1} \times \gamma^L_{\widetilde{A}_2}, \gamma^U_{\widetilde{A}_1} \times \gamma^U_{\widetilde{A}_2}]\}, \right.$$
$$\bigcup_{[\eta^L_{\widetilde{A}_1},\eta^U_{\widetilde{A}_1}]\in u_{\widetilde{A}_1},[\eta^L_{\widetilde{A}_2},\eta^U_{\widetilde{A}_2}]\in u_{\widetilde{A}_2}} \left\{\left[\sqrt[q]{(\eta^L_{\widetilde{A}_1})^q + (\eta^L_{\widetilde{A}_2})^q - (\eta^L_{\widetilde{A}_1})^q \times (\eta^L_{\widetilde{A}_2})^q}, \right.\right.$$
$$\left.\left.\left. \sqrt[q]{(\eta^U_{\widetilde{A}_1})^q + (\eta^U_{\widetilde{A}_2})^q - (\eta^U_{\widetilde{A}_1})^q \times (\eta^U_{\widetilde{A}_2})^q} \right]\right\}\right); \tag{5.16}$$

$$\lambda h_{\widetilde{A}} = \left(\bigcup_{[\gamma^L_{\widetilde{A}},\gamma^U_{\widetilde{A}}]\in u_{\widetilde{A}}} \{[\sqrt[q]{1-(1-(\gamma^L_{\widetilde{A}})^q)^\lambda}, \sqrt[q]{1-(1-(\gamma^U_{\widetilde{A}})^q)^\lambda}]\}, \right.$$
$$\left. \bigcup_{[\eta^L_{\widetilde{A}},\eta^U_{\widetilde{A}}]\in v_{\widetilde{A}}} \{[(\eta^L_{\widetilde{A}})^\lambda, (\eta^U_{\widetilde{A}})^\lambda]\} \right); \tag{5.17}$$

$$h_{\widetilde{A}}^\lambda = \left(\bigcup_{[\gamma^L_{\widetilde{A}},\gamma^U_{\widetilde{A}}]\in u_{\widetilde{A}}} \{[(\gamma^L_{\widetilde{A}})^\lambda, (\gamma^U_{\widetilde{A}})^\lambda]\}, \right.$$
$$\left. \bigcup_{[\eta^L_{\widetilde{A}},\eta^U_{\widetilde{A}}]\in v_{\widetilde{A}}} \{[\sqrt[q]{1-(1-(\eta^L_{\widetilde{A}})^q)^\lambda}, \sqrt[q]{1-(1-(\eta^U_{\widetilde{A}})^q)^\lambda}]\} \right), \tag{5.18}$$

where $q \geq 1$.

Remark 5.3 It should be mentioned that the concept of IVq-ROHFS is also referred to as the interval-valued q-rung dual hesitant fuzzy set in some references [1].

Remark 5.4 In some references (for instance, [5]), the authors use the concept "dual" in extending the notion of q-rung orthopair fuzzy set to the dual hesitant q-rung orthopair fuzzy set. This is while, the notion of "dual" is hidden in the definition of q-rung orthopair fuzzy set, and so need not to be mentioned again. Therefore, it is enough to say that only the hesitant q-rung orthopair fuzzy set, the interval-valued hesitant q-rung orthopair fuzzy set, and so on.

References

1. X. Feng, X. Shang, J. Wang, Y. Xu, A multiple attribute decision-making method based on interval-valued q-rung dual hesitant fuzzy power Hamy mean and novel score function. Comput. Appl. Math. **40**, 1–32 (2021)
2. A. Hussain, M.I. Ali, T. Mahmood, Hesitant q-rung orthopair fuzzy aggregation operators with their applications in multi-criteria decision making. Iranian J. Fuzzy Syst. **17**, 117–134 (2020)
3. D. Liu, D. Peng, Z. Liu, The distance measures between q-Rung orthopair hesitant fuzzy sets and their application in multiple criteria decision making. Int. J. Intell. Syst. **34**, 2104–2121 (2019)
4. P. Liu, P. Wang, Some q-rung orthopair fuzzy aggregation operators and their applications to multiple-attribute decision making. Int. J. Intell. Syst. **33**, 259–280 (2018)
5. J. Wang, G. Wei, C. Wei, Y. Wei, Dual hesitant q-rung orthopair fuzzy Muirhead mean operators in multiple attribute decision making. IEEE Access **7**, 67139–67166 (2019)
6. Y. Wang, Z. Shan, L. Huang, The extension of TOPSIS method for multi-attribute decision-making with q-Rung orthopair hesitant fuzzy sets. IEEE Access **8**, 165151–165167 (2020)
7. Y. Xu, X. Shang, J. Wang, K. Bai, Some interval-valued q-rung dual hesitant fuzzy Muirhead mean operators with their application to multi-attribute decision-making. IEEE Access **7**, 54724–54745 (2019)
8. R.R. Yager, Generalized orthopair fuzzy sets. IEEE Transactions on Fuzzy Systems **25**, 1222–1230 (2017)

Chapter 6
Probabilistic Hesitant Fuzzy Set

Abstract In this chapter, we firstly deal with the core of the next concepts, known as probabilistic hesitant fuzzy set. Then, the dual form of probabilistic hesitant fuzzy sets is given. We dedicate the next part of this chapter to introduce the notion of occurring probability of possible values into hesitant fuzzy linguistic term set and define probabilistic linguistic hesitant fuzzy set. On the basis of probabilistic linguistic term set, we get the definition of probabilistic linguistic dual hesitant fuzzy set by combining probabilistic dual hesitant fuzzy set with dual hesitant fuzzy linguistic term set. The other extensions of probabilistic hesitant fuzzy set including interval probabilistic hesitant fuzzy linguistic variable, probabilistic neutrosophic hesitant fuzzy set, Pythagorean probabilistic hesitant fuzzy set, and q-rung orthopair probabilistic hesitant fuzzy set will be introduced, and some their corresponding operations will be given.

6.1 Probabilistic Hesitant Fuzzy Set

If some experts assign the same value for specified alternative in a hesitant fuzzy-based decision making, then we cannot determine what their preferences are. To avoid the loss of hesitant fuzzy information in such a decision making process, Zhu [26] brought probability to the concept of hesitant fuzzy set and introduced the probabilistic hesitant fuzzy set. Zhang et al. [24] developed the probabilistic hesitant fuzzy operations and integrations to apply in decision making. Song et al. [19] represented a comparison technique for making the multiple criteria decision making more efficiently. Zhou and Xu [25] defined a probabilistic hesitant fuzzy preference relation, and then, they tested the consistency of the same for group decision making. Jiang and Ma [9] developed a number of aggregation operators under arithmetic and geometric context for probabilistic hesitant fuzzy sets.

B. Farhadinia, *Hesitant Fuzzy Set*, Computational Intelligence Methods and Applications, https://doi.org/10.1007/978-981-16-7301-6_6

Definition 6.1 ([21]) Let X be a reference set. A probabilistic hesitant fuzzy set (PHFS) A on X is defined in terms of the function $h_A(x)$ as follows:

$$A = \{\langle x, h_A(x)\rangle \mid x \in X\}, \tag{6.1}$$

where $h_A(x)$ is some different probabilistic values in [0, 1] representing the possible probabilistic membership degrees of the element $x \in X$ to the set A, respectively.

Here, for all $x \in X$, we have

$$h_A(x) = \bigcup_{\langle \gamma_A, \wp_{\gamma_A}\rangle \in h_A(x)} \{\langle \gamma_A, \wp_{\gamma_A}\rangle\}, \tag{6.2}$$

and it is considered $\gamma_A^+ \in \bigcup_{x \in X} \max\{\gamma_A(x)\}$ which implies that $0 \leq \gamma_A \leq 1$ together with $0 \leq \gamma_A^+ \leq 1$. Moreover, it holds that $\sum_{\langle \gamma_A, \wp_{\gamma_A}\rangle \in h_A(x)} \wp_{\gamma_A} = 1$ for any $0 \leq \wp_{\gamma_A} \leq 1$.

For the sake of simplicity, $h_A(x)$ is called as the probabilistic hesitant fuzzy element (PHFE).

Remark 6.1 Throughout this book, the set of all PHFSs on the reference set X is denoted by $\mathbb{PHFS}(X)$.

Example 6.1 Let $X = \{x_1, x_2\}$ be the reference set, $h_A(x_1) = \bigcup_{\langle \gamma_A, \wp_{\gamma_A}\rangle \in h_A(x_1)} \{\langle \gamma_A, \wp_{\gamma_A}\rangle\} = \{\langle 0.2, 0.7\rangle, \langle 0.5, 0.3\rangle\}$ and $h_A(x_2) = \bigcup_{\langle \gamma_A, \wp_{\gamma_A}\rangle \in h_A(x_2)} \{\langle \gamma_A, \wp_{\gamma_A}\rangle\} = \{\langle 0.3, 0.2\rangle, \langle 0.5, 0.8\rangle\}$ be the PHFEs of x_i $(i = 1, 2)$ in the set A, respectively. Thus, $h_A(x_1)$ and $h_A(x_2)$ are two PHFEs, and then A can be considered as a PHFS, i.e.,

$$A = \{\langle x_1, \{\langle 0.2, 0.7\rangle, \langle 0.5, 0.3\rangle\}\rangle, \langle x_2, \{\langle 0.3, 0.2\rangle, \langle 0.5, 0.8\rangle\}\rangle\}.$$

For the PHFEs $h_A = \bigcup_{\langle \gamma_A, \wp_{\gamma_A}\rangle \in h_A} \{\langle \gamma_A, \wp_{\gamma_A}\rangle\}$, $h_{A_1} = \bigcup_{\langle \gamma_{A_1}, \wp_{\gamma_{A_1}}\rangle \in h_{A_1}} \{\langle \gamma_{A_1}, \wp_{\gamma_{A_1}}\rangle\}$ and $h_{A_2} = \bigcup_{\langle \gamma_{A_2}, \wp_{\gamma_{A_2}}\rangle \in h_{A_2}} \{\langle \gamma_{A_2}, \wp_{\gamma_{A_2}}\rangle\}$ the following operations are defined (see [4]):

$$\begin{aligned} h_{A_1} \oplus h_{A_2} &= \bigcup_{\langle \gamma_{A_1}, \wp_{\gamma_{A_1}}\rangle \in h_{A_1}} \{\langle \gamma_{A_1}, \wp_{\gamma_{A_1}}\rangle\} \oplus \bigcup_{\langle \gamma_{A_2}, \wp_{\gamma_{A_2}}\rangle \in h_{A_2}} \{\langle \gamma_{A_2}, \wp_{\gamma_{A_2}}\rangle\} \\ &= \bigcup_{\langle \gamma_{A_1}, \wp_{\gamma_{A_1}}\rangle \in h_{A_1}, \langle \gamma_{A_2}, \wp_{\gamma_{A_2}}\rangle \in h_{A_2}} \{\langle \gamma_{A_1} + \gamma_{A_2} - \gamma_{A_1} \times \gamma_{A_2}, \\ &\quad 1 - (1 - \wp_{\gamma_{A_1}}) \times (1 - \wp_{\gamma_{A_2}})\rangle\}; \end{aligned} \tag{6.3}$$

$$h_{A_1} \otimes h_{A_2} = \bigcup_{\langle \gamma_{A_1}, \wp_{\gamma_{A_1}}\rangle \in h_{A_1}} \{\langle \gamma_{A_1}, \wp_{\gamma_{A_1}}\rangle\} \otimes \bigcup_{\langle \gamma_{A_2}, \wp_{\gamma_{A_2}}\rangle \in h_{A_2}} \{\langle \gamma_{A_2}, \wp_{\gamma_{A_2}}\rangle\}$$

$$= \bigcup_{\langle \gamma_{A_1}, \wp_{\gamma_{A_1}} \rangle \in h_{A_1}, \langle \gamma_{A_2}, \wp_{\gamma_{A_2}} \rangle \in h_{A_2}} \{\langle \gamma_{A_1} \times \gamma_{A_2}, \wp_{\gamma_{A_1}} \times \wp_{\gamma_{A_2}} \rangle\}; \quad (6.4)$$

$$\lambda h_A = \bigcup_{\langle \gamma_A, \wp_{\gamma_A} \rangle \in h_A} \{\langle 1 - (1 - \gamma_A)^\lambda, \wp_{\gamma_A} \rangle\}; \quad (6.5)$$

$$h_A{}^\lambda = \bigcup_{\langle \gamma_A, \wp_{\gamma_A} \rangle \in h_A} \{\langle (\gamma_A)^\lambda, \wp_{\gamma_A} \rangle\}. \quad (6.6)$$

6.2 Probabilistic Dual Hesitant Fuzzy Set

As a pioneer work concerning on the probabilistic dual hesitant fuzzy set, Hao et al. [8] proposed a class of operational laws and aggregation operators for probabilistic dual hesitant fuzzy sets. Then, they developed an entropy measure of probabilistic dual hesitant fuzzy sets for enhancing a visual analysis technique. In the sequel, Ren et al. [13] extended TODIM technique for probabilistic dual hesitant fuzzy sets in the application of enterprise strategic assessment. Further, Garg and Kaur [5] proposed a robust correlation coefficient for probabilistic dual hesitant fuzzy sets. Ren et al. [14] investigated a strategy selection process with an integrated AHP and VIKOR technique under probabilistic dual hesitant fuzzy set information.

Definition 6.2 ([8]) Let X be a reference set. A probabilistic dual hesitant fuzzy set (PDHFS) $\widetilde{A}$ on X is defined in terms of two functions $u_{\widetilde{A}}(x)$ and $v_{\widetilde{A}}(x)$ as follows:

$$\widetilde{A} = \{\langle x, u_{\widetilde{A}}(x), v_{\widetilde{A}}(x) \rangle \mid x \in X\}, \quad (6.7)$$

where $u_{\widetilde{A}}(x)$ and $v_{\widetilde{A}}(x)$ are some different probabilistic values in [0, 1] representing the possible probabilistic membership degrees and probabilistic non-membership degrees of the element $x \in X$ to the set $\widetilde{A}$, respectively.

Here, for all $x \in X$, if we take $u_{\widetilde{A}}(x) = \bigcup_{\langle \gamma_{\widetilde{A}}, \wp_{\gamma_{\widetilde{A}}} \rangle \in u_{\widetilde{A}}(x)} \{\langle \gamma_{\widetilde{A}}, \wp_{\gamma_{\widetilde{A}}} \rangle\}$, $v_{\widetilde{A}}(x) = \bigcup_{\langle \eta_{\widetilde{A}}, \wp_{\eta_{\widetilde{A}}} \rangle \in v_{\widetilde{A}}(x)} \{\langle \eta_{\widetilde{A}}, \wp_{\eta_{\widetilde{A}}} \rangle\}$, $\gamma_{\widetilde{A}}^+ \in \bigcup_{x \in X} \max\{\gamma_{\widetilde{A}}(x)\}$ and $\eta_{\widetilde{A}}^+ \in \bigcup_{x \in X} \max\{\eta_{\widetilde{A}}(x)\}$, then we find that

$$0 \le \gamma_{\widetilde{A}}, \eta_{\widetilde{A}} \le 1, \quad 0 \le \gamma_{\widetilde{A}}^+ + \eta_{\widetilde{A}}^+ \le 1.$$

Moreover, $\sum_{\langle \gamma_{\widetilde{A}}, \wp_{\gamma_{\widetilde{A}}} \rangle \in u_{\widetilde{A}}(x)} \wp_{\gamma_{\widetilde{A}}} = \sum_{\langle \eta_{\widetilde{A}}, \wp_{\eta_{\widetilde{A}}} \rangle \in v_{\widetilde{A}}(x)} \wp_{\eta_{\widetilde{A}}} = 1$ for any $0 \le \wp_{\gamma_{\widetilde{A}}}, \wp_{\eta_{\widetilde{A}}} \le 1$.

For the sake of simplicity, $h_{\widetilde{A}}(x) = (u_{\widetilde{A}}(x), v_{\widetilde{A}}(x))$ is called the probabilistic dual hesitant fuzzy element (PDHFE).

Remark 6.2 Throughout this book, the set of all PDHFSs on the reference set X is denoted by $\mathbb{PDHFS}(X)$.

Example 6.2 Let $X = \{x_1, x_2\}$ be the reference set, $h_{\widetilde{A}}(x_1) = (\bigcup_{\langle \gamma_{\widetilde{A}}, \wp_{\gamma_{\widetilde{A}}}\rangle \in u_{\widetilde{A}}(x_1)} \{\langle \gamma_{\widetilde{A}}, \wp_{\gamma_{\widetilde{A}}}\rangle\}, \bigcup_{\langle \eta_{\widetilde{A}}, \wp_{\eta_{\widetilde{A}}}\rangle \in v_{\widetilde{A}}(x_1)} \{\langle \eta_{\widetilde{A}}, \wp_{\eta_{\widetilde{A}}}\rangle\} = (\{\langle 0.2, 0.7\rangle, \langle 0.5, 0.3\rangle\}, \{\langle 0.7, 1\rangle\})$ and $h_{\widetilde{A}}(x_2) = (\bigcup_{\langle \gamma_{\widetilde{A}}, \wp_{\gamma_{\widetilde{A}}}\rangle \in u_{\widetilde{A}}(x_2)} \{\langle \gamma_{\widetilde{A}}, \wp_{\gamma_{\widetilde{A}}}\rangle\}, \bigcup_{\langle \eta_{\widetilde{A}}, \wp_{\eta_{\widetilde{A}}}\rangle \in v_{\widetilde{A}}(x_2)} \{\langle \eta_{\widetilde{A}}, \wp_{\eta_{\widetilde{A}}}\rangle\} = (\big\{\langle 0.3, 0.2\rangle, \langle 0.5, 0.8\rangle\big\}, \{\langle 0.2, 1\rangle\})$ be the PDHFEs of x_i $(i = 1, 2)$ in the set $\widetilde{A}$, respectively. Thus, $h_{\widetilde{A}}(x_1)$ and $h_{\widetilde{A}}(x_2)$ are two PDHFEs, and then $\widetilde{A}$ can be considered as a PDHFS, i.e.,

$$\begin{aligned}\widetilde{A} = \{&\langle x_1, (\{\langle 0.2, 0.7\rangle, \langle 0.5, 0.3\rangle\}, \{\langle 0.7, 1\rangle\})\rangle,\\ &\langle x_2, (\{\langle 0.3, 0.2\rangle, \langle 0.5, 0.8\rangle\}, \{\langle 0.2, 1\rangle\})\rangle\}.\end{aligned}$$

For the PDHFEs
$h_{\widetilde{A}} = (u_{\widetilde{A}}, v_{\widetilde{A}}) = (\bigcup_{\langle \gamma_{\widetilde{A}}, \wp_{\gamma_{\widetilde{A}}}\rangle \in u_{\widetilde{A}}} \{\langle \gamma_{\widetilde{A}}, \wp_{\gamma_{\widetilde{A}}}\rangle\}, \bigcup_{\langle \eta_{\widetilde{A}}, \wp_{\eta_{\widetilde{A}}}\rangle \in v_{\widetilde{A}}} \{\langle \eta_{\widetilde{A}}, \wp_{\eta_{\widetilde{A}}}\rangle\})$,
$h_{\widetilde{A}_1} = (u_{\widetilde{A}_1}, v_{\widetilde{A}_1}) = (\bigcup_{\langle \gamma_{\widetilde{A}_1}, \wp_{\gamma_{\widetilde{A}_1}}\rangle \in u_{\widetilde{A}_1}} \{\langle \gamma_{\widetilde{A}_1}, \wp_{\gamma_{\widetilde{A}_1}}\rangle\}, \bigcup_{\langle \eta_{\widetilde{A}_1}, \wp_{\eta_{\widetilde{A}_1}}\rangle \in v_{\widetilde{A}_1}} \{\langle \eta_{\widetilde{A}_1}, \wp_{\eta_{\widetilde{A}_1}}\rangle\})$
and
$h_{\widetilde{A}_2} = (u_{\widetilde{A}_2}, v_{\widetilde{A}_2}) = (\bigcup_{\langle \gamma_{\widetilde{A}_2}, \wp_{\gamma_{\widetilde{A}_2}}\rangle \in u_{\widetilde{A}_2}} \{\langle \gamma_{\widetilde{A}_2}, \wp_{\gamma_{\widetilde{A}_2}}\rangle\}, \bigcup_{\langle \eta_{\widetilde{A}_2}, \wp_{\eta_{\widetilde{A}_2}}\rangle \in v_{\widetilde{A}_2}} \{\langle \eta_{\widetilde{A}_2}, \wp_{\eta_{\widetilde{A}_2}}\rangle\})$
the following operations are defined (see [8]):

$$\begin{aligned}h_{\widetilde{A}_1} \oplus h_{\widetilde{A}_2} = \Bigg(&\bigcup_{\langle \gamma_{\widetilde{A}_1}, \wp_{\gamma_{\widetilde{A}_1}}\rangle \in u_{\widetilde{A}_1}, \langle \gamma_{\widetilde{A}_2}, \wp_{\gamma_{\widetilde{A}_2}}\rangle \in u_{\widetilde{A}_2}} \{\langle \gamma_{\widetilde{A}_1} + \gamma_{\widetilde{A}_2} - \gamma_{\widetilde{A}_1} \times \gamma_{\widetilde{A}_2}, \wp_{\gamma_{\widetilde{A}_1}} \times \wp_{\gamma_{\widetilde{A}_2}}\rangle\},\\ &\bigcup_{\langle \eta_{\widetilde{A}_1}, \wp_{\eta_{\widetilde{A}_1}}\rangle \in v_{\widetilde{A}_1}, \langle \eta_{\widetilde{A}_2}, \wp_{\eta_{\widetilde{A}_2}}\rangle \in v_{\widetilde{A}_2}} \{\langle \eta_{\widetilde{A}_1} \times \eta_{\widetilde{A}_2}, \wp_{\eta_{\widetilde{A}_1}} \times \wp_{\eta_{\widetilde{A}_2}}\rangle\}\Bigg);\end{aligned} \tag{6.8}$$

$$\begin{aligned}h_{\widetilde{A}_1} \otimes h_{\widetilde{A}_2} = \Bigg(&\bigcup_{\langle \gamma_{\widetilde{A}_1}, \wp_{\gamma_{\widetilde{A}_1}}\rangle \in u_{\widetilde{A}_1}, \langle \gamma_{\widetilde{A}_2}, \wp_{\gamma_{\widetilde{A}_2}}\rangle \in u_{\widetilde{A}_2}} \Big\{\langle \gamma_{\widetilde{A}_1} \times \gamma_{\widetilde{A}_2},\\ &\wp_{\gamma_{\widetilde{A}_1}} \times \wp_{\gamma_{\widetilde{A}_2}}\rangle\Big\},\\ &\bigcup_{\langle \eta_{\widetilde{A}_1}, \wp_{\eta_{\widetilde{A}_1}}\rangle \in v_{\widetilde{A}_1}, \langle \eta_{\widetilde{A}_2}, \wp_{\eta_{\widetilde{A}_2}}\rangle \in v_{\widetilde{A}_2}} \Big\{\langle \eta_{\widetilde{A}_1} + \eta_{\widetilde{A}_2} - \eta_{\widetilde{A}_1} \times \eta_{\widetilde{A}_2},\\ &\wp_{\eta_{\widetilde{A}_1}} \times \wp_{\eta_{\widetilde{A}_2}}\rangle\Big\}\Bigg);\end{aligned} \tag{6.9}$$

$$\lambda h_{\tilde{A}} = \left(\bigcup_{\langle \gamma_{\tilde{A}}, \wp_{\gamma_{\tilde{A}}} \rangle \in u_{\tilde{A}}} \{\langle 1-(1-\gamma_{\tilde{A}})^{\lambda}, \wp_{\gamma_{\tilde{A}}} \rangle\}, \bigcup_{\langle \eta_{\tilde{A}}, \wp_{\eta_{\tilde{A}}} \rangle \in v_{\tilde{A}}} \{\langle (\eta_{\tilde{A}})^{\lambda}, \wp_{\eta_{\tilde{A}}} \rangle\} \right); \quad (6.10)$$

$$h_{\tilde{A}}^{\lambda} = \left(\bigcup_{\langle \gamma_{\tilde{A}}, \wp_{\gamma_{\tilde{A}}} \rangle \in u_{\tilde{A}}} \{\langle (\gamma_{\tilde{A}})^{\lambda}, \wp_{\gamma_{\tilde{A}}} \rangle\}, \bigcup_{\langle \eta_{\tilde{A}}, \wp_{\eta_{\tilde{A}}} \rangle \in v_{\tilde{A}}} \{\langle 1-(1-\eta_{\tilde{A}})^{\lambda}, \wp_{\eta_{\tilde{A}}} \rangle\} \right). \quad (6.11)$$

Remark 6.3 It should be mentioned that the concept of PDHFS is also referred to as the dual hesitant fuzzy probability set [3], and the probabilistic intuitionistic hesitant fuzzy set [23] in some references.

6.3 Probabilistic Linguistic Hesitant Fuzzy Set

Gong et al. [7] presented the probabilistic linguistic hesitant fuzzy preference relation on the basis of probabilistic linguistic hesitant fuzzy set which not only provides flexible linguistic expression for decision makers but also gives the occurrence probability of each element in the probabilistic linguistic hesitant fuzzy preference relation. Zhao and Huang [22] developed the notions of score and variance functions for probabilistic linguistic hesitant fuzzy sets together with a set of basic operations. Then, they implemented a probabilistic linguistic hesitant fuzzy-based distance measure in multiple criteria decision making.

Definition 6.3 ([10]) Let X be a reference set, and $\mathfrak{S} = \{s_\alpha \mid \alpha = 0, 1, \ldots, \tau\}$ be a linguistic term set. A probabilistic linguistic hesitant fuzzy set (PLHFS) on X is mathematically shown in terms of

$$\mathbb{A}^{\mathfrak{S}} = \{\langle x, h_{\mathbb{A}^{\mathfrak{S}}}(x) \rangle \mid x \in X\}. \quad (6.12)$$

Here, $h_{\mathbb{A}^{\mathfrak{S}}}(x)$ is a set of some possible pairs of linguistic term set $\mathfrak{S}$ together with probabilistic values which can be characterized in the form of

$$h_{\mathbb{A}^{\mathfrak{S}}}(x) = \langle s_\alpha(x), \bigcup_{\langle \gamma_{\mathbb{A}^{\mathfrak{S}}}, \wp_{\gamma_{\mathbb{A}^{\mathfrak{S}}}} \rangle \in h_{\mathbb{A}^{\mathfrak{S}}}(x)} \{\langle \gamma_{\mathbb{A}^{\mathfrak{S}}}, \wp_{\gamma_{\mathbb{A}^{\mathfrak{S}}}} \rangle\} \rangle, \quad (6.13)$$

where $s_\alpha(x) \in \mathfrak{S}$ denotes the linguistic terms involved in $h_{\mathbb{A}^{\mathfrak{S}}}(x)$.

The notation $h_{\mathbb{A}^{\mathfrak{S}}}(x)$ is called hereafter probabilistic linguistic hesitant fuzzy set (PLHFE) on X.

Remark 6.4 Throughout this book, the set of all PLHFSs on the reference set X is denoted by $\mathbb{PLHFS}(X)$.

Example 6.3 Let $X = \{x_1, x_2\}$ be the reference set, $h_{\mathbb{A}^{\mathfrak{S}}}(x_1) = \langle s_2, \{\langle 0.2, 0.3\rangle, \langle 0.4, 0.6\rangle, \langle 0.5, 0.1\rangle\}\rangle$ and $h_{\mathbb{A}^{\mathfrak{S}}}(x_2) = \langle s_1, \{\langle 0.3, 0.5\rangle, \langle 0.4, 0.5\rangle\}\rangle$ be the PLHFEs of x $(i = 1, 2)$ to a set $\mathbb{A}^{\mathfrak{S}}$, respectively. Then $\mathbb{A}^{\mathfrak{S}}$ can be considered as a PLHFS, i.e.,

$$\mathbb{A}^{\mathfrak{S}} = \{\langle x_1, \langle s_2, \{\langle 0.2, 0.3\rangle, \langle 0.4, 0.6\rangle, \langle 0.5, 0.1\rangle\}\rangle\rangle,\\ \langle x_2, \langle s_1, \{\langle 0.3, 0.5\rangle, \langle 0.4, 0.5\rangle\}\rangle\rangle\}.$$

Given three PLHFEs represented by $h_{\mathbb{A}^{\mathfrak{S}}} = \langle s_\alpha, \bigcup_{\langle \gamma_{\mathbb{A}^{\mathfrak{S}}}, \wp_{\gamma_{\mathbb{A}^{\mathfrak{S}}}}\rangle \in h_{\mathbb{A}^{\mathfrak{S}}}}\{\langle \gamma_{\mathbb{A}^{\mathfrak{S}}}, \wp_{\gamma_{\mathbb{A}^{\mathfrak{S}}}}\rangle\}\rangle$, $h_{\mathbb{A}_1^{\mathfrak{S}}} = \langle s_\alpha, \bigcup_{\langle \gamma_{\mathbb{A}_1^{\mathfrak{S}}}, \wp_{\gamma_{\mathbb{A}_1^{\mathfrak{S}}}}\rangle \in h_{\mathbb{A}_1^{\mathfrak{S}}}}\{\langle \gamma_{\mathbb{A}_1^{\mathfrak{S}}}, \wp_{\gamma_{\mathbb{A}_1^{\mathfrak{S}}}}\rangle\}\rangle$, and
$h_{\mathbb{A}_2^{\mathfrak{S}}} = \langle s_\alpha, \bigcup_{\langle \gamma_{\mathbb{A}_2^{\mathfrak{S}}}, \wp_{\gamma_{\mathbb{A}_2^{\mathfrak{S}}}}\rangle \in h_{\mathbb{A}_2^{\mathfrak{S}}}}\{\langle \gamma_{\mathbb{A}_2^{\mathfrak{S}}}, \wp_{\gamma_{\mathbb{A}_2^{\mathfrak{S}}}}\rangle\}\rangle$, some set and arithmetic operations on the PLHFEs, which are also PLHFEs, can be described as follows (see, e.g., [10]):

$$h^c_{\mathbb{A}^{\mathfrak{S}}} = \langle s_{\tau-\alpha}, \bigcup_{\langle \gamma_{\mathbb{A}^{\mathfrak{S}}}, \wp_{\gamma_{\mathbb{A}^{\mathfrak{S}}}}\rangle \in h_{\mathbb{A}^{\mathfrak{S}}}} \{\langle 1-\gamma_{\mathbb{A}^{\mathfrak{S}}}, \wp_{\gamma_{\mathbb{A}^{\mathfrak{S}}}}\rangle\}\rangle, \tag{6.14}$$

$$h_{\mathbb{A}_1^{\mathfrak{S}}} \cup h_{\mathbb{A}_2^{\mathfrak{S}}}\\ = \left\langle s_{\max\{\alpha_1,\alpha_2\}}, \bigcup_{\langle \gamma_{\mathbb{A}_1^{\mathfrak{S}}}, \wp_{\gamma_{\mathbb{A}_1^{\mathfrak{S}}}}\rangle \in h_{\mathbb{A}_1^{\mathfrak{S}}}, \langle \gamma_{\mathbb{A}_2^{\mathfrak{S}}}, \wp_{\gamma_{\mathbb{A}_2^{\mathfrak{S}}}}\rangle \in h_{\mathbb{A}_2^{\mathfrak{S}}}} \Big\{\langle \max\{\gamma_{\mathbb{A}_1^{\mathfrak{S}}}, \gamma_{\mathbb{A}_2^{\mathfrak{S}}}\},\\ \max\{\wp_{\gamma_{\mathbb{A}_1^{\mathfrak{S}}}}, \wp_{\gamma_{\mathbb{A}_2^{\mathfrak{S}}}}\}\rangle\Big\}\right\rangle; \tag{6.15}$$

$$h_{\mathbb{A}_1^{\mathfrak{S}}} \cap h_{\mathbb{A}_2^{\mathfrak{S}}}\\ = \left\langle s_{\min\{\alpha_1,\alpha_2\}}, \bigcup_{\langle \gamma_{\mathbb{A}_1^{\mathfrak{S}}}, \wp_{\gamma_{\mathbb{A}_1^{\mathfrak{S}}}}\rangle \in h_{\mathbb{A}_1^{\mathfrak{S}}}, \langle \gamma_{\mathbb{A}_2^{\mathfrak{S}}}, \wp_{\gamma_{\mathbb{A}_2^{\mathfrak{S}}}}\rangle \in h_{\mathbb{A}_2^{\mathfrak{S}}}} \Big\{\langle \min\{\gamma_{\mathbb{A}_1^{\mathfrak{S}}}, \gamma_{\mathbb{A}_2^{\mathfrak{S}}}\},\\ \min\{\wp_{\gamma_{\mathbb{A}_1^{\mathfrak{S}}}}, \wp_{\gamma_{\mathbb{A}_2^{\mathfrak{S}}}}\}\rangle\Big\}\right\rangle; \tag{6.16}$$

$$h_{\mathbb{A}_1^{\mathfrak{S}}} \oplus h_{\mathbb{A}_2^{\mathfrak{S}}}\\ = \left\langle s_{\alpha_1+\alpha_2}, \bigcup_{\langle \gamma_{\mathbb{A}_1^{\mathfrak{S}}}, \wp_{\gamma_{\mathbb{A}_1^{\mathfrak{S}}}}\rangle \in h_{\mathbb{A}_1^{\mathfrak{S}}}, \langle \gamma_{\mathbb{A}_2^{\mathfrak{S}}}, \wp_{\gamma_{\mathbb{A}_2^{\mathfrak{S}}}}\rangle \in h_{\mathbb{A}_2^{\mathfrak{S}}}} \right.$$

$$\left\{\langle\gamma_{\mathbf{A}_1^{\mathfrak{S}}} + \gamma_{\mathbf{A}_2^{\mathfrak{S}}} - \gamma_{\mathbf{A}_1^{\mathfrak{S}}} \times \gamma_{\mathbf{A}_2^{\mathfrak{S}}}, \wp_{\gamma_{\mathbf{A}_1^{\mathfrak{S}}}} \times \wp_{\gamma_{\mathbf{A}_2^{\mathfrak{S}}}}\rangle\right\}\Big\rangle; \tag{6.17}$$

$$h_{\mathbf{A}_1^{\mathfrak{S}}} \otimes h_{\mathbf{A}_2^{\mathfrak{S}}} = \left\langle s_{\alpha_1 \times \alpha_2}, \bigcup_{\langle\gamma_{\mathbf{A}_1^{\mathfrak{S}}}, \wp_{\gamma_{\mathbf{A}_1^{\mathfrak{S}}}}\rangle \in h_{\mathbf{A}_1^{\mathfrak{S}}}, \langle\gamma_{\mathbf{A}_2^{\mathfrak{S}}}, \wp_{\gamma_{\mathbf{A}_2^{\mathfrak{S}}}}\rangle \in h_{\mathbf{A}_2^{\mathfrak{S}}}} \left\{\langle\gamma_{\mathbf{A}_1^{\mathfrak{S}}} \times \gamma_{\mathbf{A}_2^{\mathfrak{S}}}, \wp_{\gamma_{\mathbf{A}_1^{\mathfrak{S}}}} \times \wp_{\gamma_{\mathbf{A}_2^{\mathfrak{S}}}}\rangle\right\}\right\rangle; \tag{6.18}$$

$$h_{\mathbf{A}^{\mathfrak{S}}}{}^{\lambda} = \left\langle s_{(\alpha)^{\lambda}}, \bigcup_{\langle\gamma_{\mathbf{A}^{\mathfrak{S}}}, \wp_{\gamma_{\mathbf{A}^{\mathfrak{S}}}}\rangle \in h_{\mathbf{A}^{\mathfrak{S}}}} \{\langle(\gamma_{\mathbf{A}^{\mathfrak{S}}})^{\lambda}, \wp_{\gamma_{\mathbf{A}^{\mathfrak{S}}}}\rangle\}\right\rangle, \quad \lambda > 0; \tag{6.19}$$

$$\lambda h_{\mathbf{A}^{\mathfrak{S}}} = \left\langle s_{\lambda\alpha}, \bigcup_{\langle\gamma_{\mathbf{A}^{\mathfrak{S}}}, \wp_{\gamma_{\mathbf{A}^{\mathfrak{S}}}}\rangle \in h_{\mathbf{A}^{\mathfrak{S}}}} \{\langle 1 - (1 - \gamma_{\mathbf{A}^{\mathfrak{S}}})^{\lambda}, \wp_{\gamma_{\mathbf{A}^{\mathfrak{S}}}}\rangle\}\right\rangle, \quad \lambda > 0. \tag{6.20}$$

Remark that the probability part of union and intersection operations should be unified after operating to make the resulted PLHFEs normalized.

6.4 Probabilistic Linguistic Dual Hesitant Fuzzy Set

Based on the probabilistic linguistic term set proposed by Pang et al. [12], Gong and Chen [6] gave the definition of probabilistic linguistic dual hesitant fuzzy set by combining the concept of probabilistic dual hesitant fuzzy set with the concept of dual hesitant fuzzy linguistic term set.

Definition 6.4 ([6]) Let X be a reference set and $\mathfrak{S} = \{s_\alpha \mid \alpha = 0, 1, \ldots, \tau\}$ be a linguistic term set. A probabilistic linguistic dual hesitant fuzzy set (PLDHFS) $\widetilde{\mathbf{A}}^{\mathfrak{S}}$ on X is defined in terms of two functions $u_{\widetilde{\mathbf{A}}^{\mathfrak{S}}}(x)$ and $v_{\widetilde{\mathbf{A}}^{\mathfrak{S}}}(x)$ as follows:

$$\widetilde{\mathbf{A}}^{\mathfrak{S}} = \{\langle x, s_\alpha(x), u_{\widetilde{\mathbf{A}}^{\mathfrak{S}}}(x), v_{\widetilde{\mathbf{A}}^{\mathfrak{S}}}(x)\rangle \mid x \in X\}, \tag{6.21}$$

where $u_{\widetilde{\mathbf{A}}^{\mathfrak{S}}}(x)$ and $v_{\widetilde{\mathbf{A}}^{\mathfrak{S}}}(x)$ are some different probabilistic values in $[0, 1]$ representing the possible probabilistic membership degrees and probabilistic non-membership degrees of the element $x \in X$ to the set $\widetilde{\mathbf{A}}^{\mathfrak{S}}$, respectively. Furthermore, $s_\alpha(x) \in \mathfrak{S}$ denotes the linguistic terms involved in $\widetilde{\mathbf{A}}^{\mathfrak{S}}(x)$.

Here, for all $x \in X$, if we consider $u_{\widetilde{\mathbf{A}}^{\mathfrak{S}}}(x) = \bigcup_{\langle \gamma_{\widetilde{\mathbf{A}}^{\mathfrak{S}}}, \wp_{\gamma_{\widetilde{\mathbf{A}}^{\mathfrak{S}}}} \rangle \in u_{\widetilde{\mathbf{A}}^{\mathfrak{S}}}(x)} \{\langle \gamma_{\widetilde{\mathbf{A}}^{\mathfrak{S}}}, \wp_{\gamma_{\widetilde{\mathbf{A}}^{\mathfrak{S}}}} \rangle\}$, $v_{\widetilde{\mathbf{A}}^{\mathfrak{S}}}(x) = \bigcup_{\langle \eta_{\widetilde{\mathbf{A}}^{\mathfrak{S}}}, \wp_{\eta_{\widetilde{\mathbf{A}}^{\mathfrak{S}}}} \rangle \in v_{\widetilde{\mathbf{A}}^{\mathfrak{S}}}(x)} \{\langle \eta_{\widetilde{\mathbf{A}}^{\mathfrak{S}}}, \wp_{\eta_{\widetilde{\mathbf{A}}^{\mathfrak{S}}}} \rangle\}$, $\gamma_{\widetilde{\mathbf{A}}^{\mathfrak{S}}}{}^{+} \in \bigcup_{x \in X} \max\{\gamma_{\widetilde{\mathbf{A}}^{\mathfrak{S}}}(x)\}$ and $\eta_{\widetilde{\mathbf{A}}^{\mathfrak{S}}}{}^{+} \in \bigcup_{x \in X} \max\{\eta_{\widetilde{\mathbf{A}}^{\mathfrak{S}}}(x)\}$, then we conclude that

$$0 \le \gamma_{\widetilde{\mathbf{A}}^{\mathfrak{S}}}, \eta_{\widetilde{\mathbf{A}}^{\mathfrak{S}}} \le 1, \quad 0 \le \gamma_{\widetilde{\mathbf{A}}^{\mathfrak{S}}}{}^{+} + \eta_{\widetilde{\mathbf{A}}^{\mathfrak{S}}}{}^{+} \le 1.$$

Moreover, we have $\sum_{\langle \gamma_{\widetilde{\mathbf{A}}^{\mathfrak{S}}}, \wp_{\gamma_{\widetilde{\mathbf{A}}^{\mathfrak{S}}}} \rangle \in u_{\widetilde{\mathbf{A}}^{\mathfrak{S}}}(x)} \wp_{\gamma_{\widetilde{\mathbf{A}}^{\mathfrak{S}}}} = \sum_{\langle \eta_{\widetilde{\mathbf{A}}^{\mathfrak{S}}}, \wp_{\eta_{\widetilde{\mathbf{A}}^{\mathfrak{S}}}} \rangle \in v_{\widetilde{\mathbf{A}}^{\mathfrak{S}}}(x)} \wp_{\eta_{\widetilde{\mathbf{A}}^{\mathfrak{S}}}} = 1$ for any $0 \le \wp_{\gamma_{\widetilde{\mathbf{A}}^{\mathfrak{S}}}}, \wp_{\eta_{\widetilde{\mathbf{A}}^{\mathfrak{S}}}} \le 1$.

For the sake of simplicity, $h_{\widetilde{\mathbf{A}}^{\mathfrak{S}}}(x) = (s_\alpha(x), u_{\widetilde{\mathbf{A}}^{\mathfrak{S}}}(x), v_{\widetilde{\mathbf{A}}^{\mathfrak{S}}}(x))$ is called the probabilistic linguistic dual hesitant fuzzy element (PLDHFE).

Remark 6.5 Throughout this book, the set of all PLDHFSs on the reference set X is denoted by $\mathbb{PLDHFS}(X)$.

Example 6.4 Let $X = \{x_1, x_2\}$ be the reference set and $\mathfrak{S} = \{s_\alpha \mid \alpha = 0, 1, \ldots, 5\}$. We suppose that
$h_{\widetilde{\mathbf{A}}^{\mathfrak{S}}}(x_1) = (s_{\alpha_1}, \bigcup_{\langle \gamma_{\widetilde{\mathbf{A}}^{\mathfrak{S}}}, \wp_{\gamma_{\widetilde{\mathbf{A}}^{\mathfrak{S}}}} \rangle \in u_{\widetilde{\mathbf{A}}^{\mathfrak{S}}}(x_1)} \{\langle \gamma_{\widetilde{\mathbf{A}}^{\mathfrak{S}}}, \wp_{\gamma_{\widetilde{\mathbf{A}}^{\mathfrak{S}}}} \rangle\}, \bigcup_{\langle \eta_{\widetilde{\mathbf{A}}^{\mathfrak{S}}}, \wp_{\eta_{\widetilde{\mathbf{A}}^{\mathfrak{S}}}} \rangle \in v_{\widetilde{\mathbf{A}}^{\mathfrak{S}}}(x_1)}$ $\{\langle \eta_{\widetilde{\mathbf{A}}^{\mathfrak{S}}}, \wp_{\eta_{\widetilde{\mathbf{A}}^{\mathfrak{S}}}} \rangle\}) = (s_1, \{\langle 0.2, 0.7 \rangle, \langle 0.5, 0.3 \rangle\}, \{\langle 0.7, 1 \rangle\})$ and
$h_{\widetilde{\mathbf{A}}^{\mathfrak{S}}}(x_2) = (s_{\alpha_2}, \bigcup_{\langle \gamma_{\widetilde{\mathbf{A}}^{\mathfrak{S}}}, \wp_{\gamma_{\widetilde{\mathbf{A}}^{\mathfrak{S}}}} \rangle \in u_{\widetilde{\mathbf{A}}^{\mathfrak{S}}}(x_2)} \{\langle \gamma_{\widetilde{\mathbf{A}}^{\mathfrak{S}}}, \wp_{\gamma_{\widetilde{\mathbf{A}}^{\mathfrak{S}}}} \rangle\}$,
$\bigcup_{\langle \eta_{\widetilde{\mathbf{A}}^{\mathfrak{S}}}, \wp_{\eta_{\widetilde{\mathbf{A}}^{\mathfrak{S}}}} \rangle \in v_{\widetilde{\mathbf{A}}^{\mathfrak{S}}}(x_2)} \{\langle \eta_{\widetilde{\mathbf{A}}^{\mathfrak{S}}}, \wp_{\eta_{\widetilde{\mathbf{A}}^{\mathfrak{S}}}} \rangle\})$
$= (s_3, \{\langle 0.3, 0.2 \rangle, \langle 0.5, 0.8 \rangle\}, \{\langle 0.2, 1 \rangle\})$ are two PLDHFEs of x_i $(i = 1, 2)$ in the set $\widetilde{\mathbf{A}}^{\mathfrak{S}}$, respectively. Thus, $h_{\widetilde{\mathbf{A}}^{\mathfrak{S}}}(x_1)$ and $h_{\widetilde{\mathbf{A}}^{\mathfrak{S}}}(x_2)$ are two PLDHFEs, and then $\widetilde{\mathbf{A}}^{\mathfrak{S}}$ can be considered as a PLDHFS, i.e.,

$$\begin{aligned} \widetilde{\mathbf{A}}^{\mathfrak{S}} = \{&\langle x_1, (s_1, \{\langle 0.2, 0.7 \rangle, \langle 0.5, 0.3 \rangle\}, \{\langle 0.7, 1 \rangle\}) \rangle, \\ &\langle x_2, (s_3, \{\langle 0.3, 0.2 \rangle, \langle 0.5, 0.8 \rangle\}, \{\langle 0.2, 1 \rangle\}) \rangle\}. \end{aligned}$$

For the PLDHFEs
$h_{\widetilde{\mathbf{A}}^{\mathfrak{S}}} = (s_\alpha, u_{\widetilde{\mathbf{A}}^{\mathfrak{S}}}, v_{\widetilde{\mathbf{A}}^{\mathfrak{S}}})$
$= (s_\alpha, \bigcup_{\langle \gamma_{\widetilde{\mathbf{A}}^{\mathfrak{S}}}, \wp_{\gamma_{\widetilde{\mathbf{A}}^{\mathfrak{S}}}} \rangle \in u_{\widetilde{\mathbf{A}}^{\mathfrak{S}}}} \{\langle \gamma_{\widetilde{\mathbf{A}}^{\mathfrak{S}}}, \wp_{\gamma_{\widetilde{\mathbf{A}}^{\mathfrak{S}}}} \rangle\}, \bigcup_{\langle \eta_{\widetilde{\mathbf{A}}^{\mathfrak{S}}}, \wp_{\eta_{\widetilde{\mathbf{A}}^{\mathfrak{S}}}} \rangle \in v_{\widetilde{\mathbf{A}}^{\mathfrak{S}}}} \{\langle \eta_{\widetilde{\mathbf{A}}^{\mathfrak{S}}}, \wp_{\eta_{\widetilde{\mathbf{A}}^{\mathfrak{S}}}} \rangle\})$,
$h_{\widetilde{\mathbf{A}}_1^{\mathfrak{S}}} = (s_{\alpha_1}, u_{\widetilde{\mathbf{A}}_1^{\mathfrak{S}}}, v_{\widetilde{\mathbf{A}}_1^{\mathfrak{S}}})$
$= (s_{\alpha_1}, \bigcup_{\langle \gamma_{\widetilde{\mathbf{A}}_1^{\mathfrak{S}}}, \wp_{\gamma_{\widetilde{\mathbf{A}}_1^{\mathfrak{S}}}} \rangle \in u_{\widetilde{\mathbf{A}}_1^{\mathfrak{S}}}} \{\langle \gamma_{\widetilde{\mathbf{A}}_1^{\mathfrak{S}}}, \wp_{\gamma_{\widetilde{\mathbf{A}}_1^{\mathfrak{S}}}} \rangle\}, \bigcup_{\langle \eta_{\widetilde{\mathbf{A}}_1^{\mathfrak{S}}}, \wp_{\eta_{\widetilde{\mathbf{A}}_1^{\mathfrak{S}}}} \rangle \in v_{\widetilde{\mathbf{A}}_1^{\mathfrak{S}}}} \{\langle \eta_{\widetilde{\mathbf{A}}_1^{\mathfrak{S}}}, \wp_{\eta_{\widetilde{\mathbf{A}}_1^{\mathfrak{S}}}} \rangle\})$
and
$h_{\widetilde{\mathbf{A}}_2^{\mathfrak{S}}} = (s_{\alpha_2}, u_{\widetilde{\mathbf{A}}_2^{\mathfrak{S}}}, v_{\widetilde{\mathbf{A}}_2^{\mathfrak{S}}})$
$= (s_{\alpha_2}, \bigcup_{\langle \gamma_{\widetilde{\mathbf{A}}_2^{\mathfrak{S}}}, \wp_{\gamma_{\widetilde{\mathbf{A}}_2^{\mathfrak{S}}}} \rangle \in u_{\widetilde{\mathbf{A}}_2^{\mathfrak{S}}}} \{\langle \gamma_{\widetilde{\mathbf{A}}_2^{\mathfrak{S}}}, \wp_{\gamma_{\widetilde{\mathbf{A}}_2^{\mathfrak{S}}}} \rangle\}, \bigcup_{\langle \eta_{\widetilde{\mathbf{A}}_2^{\mathfrak{S}}}, \wp_{\eta_{\widetilde{\mathbf{A}}_2^{\mathfrak{S}}}} \rangle \in v_{\widetilde{\mathbf{A}}_2^{\mathfrak{S}}}} \{\langle \eta_{\widetilde{\mathbf{A}}_2^{\mathfrak{S}}}, \wp_{\eta_{\widetilde{\mathbf{A}}_2^{\mathfrak{S}}}} \rangle\})$
the following operations are defined (see [6]):

$$\begin{aligned} h_{\widetilde{\mathbf{A}}_1^{\mathfrak{S}}} \oplus h_{\widetilde{\mathbf{A}}_2^{\mathfrak{S}}} = \Big(&s_{\alpha_1+\alpha_2}, \\ &\bigcup_{\langle \gamma_{\widetilde{\mathbf{A}}_1^{\mathfrak{S}}}, \wp_{\gamma_{\widetilde{\mathbf{A}}_1^{\mathfrak{S}}}} \rangle \in u_{\widetilde{\mathbf{A}}_1^{\mathfrak{S}}}, \langle \gamma_{\widetilde{\mathbf{A}}_2^{\mathfrak{S}}}, \wp_{\gamma_{\widetilde{\mathbf{A}}_2^{\mathfrak{S}}}} \rangle \in u_{\widetilde{\mathbf{A}}_2^{\mathfrak{S}}}} \Big\{\langle \gamma_{\widetilde{\mathbf{A}}_1^{\mathfrak{S}}} + \gamma_{\widetilde{\mathbf{A}}_2^{\mathfrak{S}}} - \gamma_{\widetilde{\mathbf{A}}_1^{\mathfrak{S}}} \times \gamma_{\widetilde{\mathbf{A}}_2^{\mathfrak{S}}}, \end{aligned}$$

$$\left.\wp\gamma_{\widetilde{\mathbf{A}}_1}\mathfrak{S} \times \wp\gamma_{\widetilde{\mathbf{A}}_2}\mathfrak{S}\rangle\right\},$$

$$\left.\bigcup_{\langle\eta_{\widetilde{\mathbf{A}}_1}\mathfrak{S},\wp\eta_{\widetilde{\mathbf{A}}_1}\mathfrak{S}\rangle\in v_{\widetilde{\mathbf{A}}_1}\mathfrak{S},\langle\eta_{\widetilde{\mathbf{A}}_2}\mathfrak{S},\wp\eta_{\widetilde{\mathbf{A}}_2}\mathfrak{S}\rangle\in v_{\widetilde{\mathbf{A}}_2}\mathfrak{S}} \{\langle\eta_{\widetilde{\mathbf{A}}_1}\mathfrak{S} \times \eta_{\widetilde{\mathbf{A}}_2}\mathfrak{S}, \wp\eta_{\widetilde{\mathbf{A}}_1}\mathfrak{S} \times \wp\eta_{\widetilde{\mathbf{A}}_2}\mathfrak{S}\rangle\}\right); \tag{6.22}$$

$$h_{\widetilde{\mathbf{A}}_1}\mathfrak{S} \otimes h_{\widetilde{\mathbf{A}}_2}\mathfrak{S} = \Big(s_{\alpha_1\times\alpha_2},$$

$$\bigcup_{\langle\gamma_{\widetilde{\mathbf{A}}_1}\mathfrak{S},\wp\gamma_{\widetilde{\mathbf{A}}_1}\mathfrak{S}\rangle\in u_{\widetilde{\mathbf{A}}_1}\mathfrak{S},\langle\gamma_{\widetilde{\mathbf{A}}_2}\mathfrak{S},\wp\gamma_{\widetilde{\mathbf{A}}_2}\mathfrak{S}\rangle\in u_{\widetilde{\mathbf{A}}_2}\mathfrak{S}} \{\langle\gamma_{\widetilde{\mathbf{A}}_1}\mathfrak{S} \times \gamma_{\widetilde{\mathbf{A}}_2}\mathfrak{S}, \wp\gamma_{\widetilde{\mathbf{A}}_1}\mathfrak{S} \times \wp\gamma_{\widetilde{\mathbf{A}}_2}\mathfrak{S}\rangle\},$$

$$\bigcup_{\langle\eta_{\widetilde{\mathbf{A}}_1}\mathfrak{S},\wp\eta_{\widetilde{\mathbf{A}}_1}\mathfrak{S}\rangle\in v_{\widetilde{\mathbf{A}}_1}\mathfrak{S},\langle\eta_{\widetilde{\mathbf{A}}_2}\mathfrak{S},\wp\eta_{\widetilde{\mathbf{A}}_2}\mathfrak{S}\rangle\in v_{\widetilde{\mathbf{A}}_2}\mathfrak{S}} \Big\{\langle\eta_{\widetilde{\mathbf{A}}_1}\mathfrak{S} + \eta_{\widetilde{\mathbf{A}}_2}\mathfrak{S} - \eta_{\widetilde{\mathbf{A}}_1}\mathfrak{S} \times \eta_{\widetilde{\mathbf{A}}_2}\mathfrak{S},$$

$$\wp\eta_{\widetilde{\mathbf{A}}_1}\mathfrak{S} \times \wp\eta_{\widetilde{\mathbf{A}}_2}\mathfrak{S}\rangle\Big\}\Big); \tag{6.23}$$

$$\lambda h_{\widetilde{\mathbf{A}}}\mathfrak{S} = \left(s_{\lambda\alpha}, \bigcup_{\langle\gamma_{\widetilde{\mathbf{A}}}\mathfrak{S},\wp\gamma_{\widetilde{\mathbf{A}}}\mathfrak{S}\rangle\in u_{\widetilde{\mathbf{A}}}\mathfrak{S}} \{\langle 1-(1-\gamma_{\widetilde{\mathbf{A}}}\mathfrak{S})^{\lambda}, \wp\gamma_{\widetilde{\mathbf{A}}}\mathfrak{S}\rangle\},\right.$$

$$\left.\bigcup_{\langle\eta_{\widetilde{\mathbf{A}}}\mathfrak{S},\wp\eta_{\widetilde{\mathbf{A}}}\mathfrak{S}\rangle\in v_{\widetilde{\mathbf{A}}}\mathfrak{S}} \{\langle(\eta_{\widetilde{\mathbf{A}}}\mathfrak{S})^{\lambda}, \wp\eta_{\widetilde{\mathbf{A}}}\mathfrak{S}\rangle\}\right); \tag{6.24}$$

$$h_{\widetilde{\mathbf{A}}}\mathfrak{S}^{\lambda} = \Big(s_{(\alpha)^{\lambda}},$$

$$\bigcup_{\langle\gamma_{\widetilde{\mathbf{A}}}\mathfrak{S},\wp\gamma_{\widetilde{\mathbf{A}}}\mathfrak{S}\rangle\in u_{\widetilde{\mathbf{A}}}\mathfrak{S}} \{\langle(\gamma_{\widetilde{\mathbf{A}}}\mathfrak{S})^{\lambda}, \wp\gamma_{\widetilde{\mathbf{A}}}\mathfrak{S}\rangle\},$$

$$\left.\bigcup_{\langle\eta_{\widetilde{\mathbf{A}}}\mathfrak{S},\wp\eta_{\widetilde{\mathbf{A}}}\mathfrak{S}\rangle\in v_{\widetilde{\mathbf{A}}}\mathfrak{S}} \{\langle 1-(1-\eta_{\widetilde{\mathbf{A}}}\mathfrak{S})^{\lambda}, \wp\eta_{\widetilde{\mathbf{A}}}\mathfrak{S}\rangle\}\right). \tag{6.25}$$

6.5 Interval Probability Hesitant Fuzzy Linguistic Variable

Xian et al. [20] introduced the interval probability hesitant fuzzy linguistic variable by taking the concept of hesitant fuzzy linguistic term set as the evaluation part together with a novel element-reliability of evaluation. They also gave the definition of operation rules and comparison techniques for interval probability hesitant fuzzy linguistic variables.

Definition 6.5 ([20]) Let X be a reference set, and $\mathfrak{S} = \{s_\alpha \mid \alpha = -\tau, \ldots, -1, 0, 1, \ldots, \tau\}$ be a linguistic term set. An interval probability hesitant fuzzy linguistic variable (IPHFLV) on X is mathematically shown in terms of

$$\underline{A}^{\mathfrak{S}} = \{\langle x, h_{\underline{A}^{\mathfrak{S}}}(x)\rangle \mid x \in X\}. \tag{6.26}$$

Here, $h_{\underline{A}^{\mathfrak{S}}}(x)$ is a set of some possible pairs of linguistic term set $\mathfrak{S}$ together with probabilistic values which can be characterized by

$$h_{\underline{A}^{\mathfrak{S}}}(x) = \bigcup_{\langle s_\alpha, [\wp^L_{\gamma_{\underline{A}^{\mathfrak{S}}}}, \wp^U_{\gamma_{\underline{A}^{\mathfrak{S}}}}]\rangle \in h_{\underline{A}^{\mathfrak{S}}}(x)} \langle \{s_\alpha\}, [\wp^L_{\gamma_{\underline{A}^{\mathfrak{S}}}}, \wp^U_{\gamma_{\underline{A}^{\mathfrak{S}}}}]\rangle, \tag{6.27}$$

where $s_\alpha \in \mathfrak{S}$ denotes the linguistic terms involved in $h_{\underline{A}^{\mathfrak{S}}}(x)$.

Remark 6.6 Throughout this book, the set of all IPHFLVs on the reference set X is denoted by $\mathbb{IPHFLV}(X)$.

Example 6.5 Let $X = \{x_1, x_2\}$ be the reference set, $h_{\underline{A}^{\mathfrak{S}}}(x_1) = \langle \{s_{-1}, s_0, s_1\}, [0.6, 0.8]\rangle$ and $h_{\underline{A}^{\mathfrak{S}}}(x_2) = \langle \{s_0, s_1, s_2\}, [0.7, 0.8]\rangle$ be the IPHFLVs of x_i $(i = 1, 2)$ to a set $\underline{A}^{\mathfrak{S}}$, respectively. Then $\underline{A}^{\mathfrak{S}}$ can be considered as a IPHFLV, i.e.,

$$\underline{A}^{\mathfrak{S}} = \{\langle x_1, \langle \{s_{-1}, s_0, s_1\}, [0.6, 0.8]\rangle\rangle, \ \langle x_2, \langle \{s_0, s_1, s_2\}, [0.7, 0.8]\rangle\rangle\}.$$

Given three IPHFLVs represented by
$h_{\underline{A}^{\mathfrak{S}}} = \bigcup_{\langle s_\alpha, [\wp^L_{\gamma_{\underline{A}^{\mathfrak{S}}}}, \wp^U_{\gamma_{\underline{A}^{\mathfrak{S}}}}]\rangle \in h_{\underline{A}^{\mathfrak{S}}}} \langle \{s_\alpha\}, [\wp^L_{\gamma_{\underline{A}^{\mathfrak{S}}}}, \wp^U_{\gamma_{\underline{A}^{\mathfrak{S}}}}]\rangle$,
$h_{\underline{A_1}^{\mathfrak{S}}} = \bigcup_{\langle s_{\alpha_1}, [\wp^L_{\gamma_{\underline{A_1}^{\mathfrak{S}}}}, \wp^U_{\gamma_{\underline{A_1}^{\mathfrak{S}}}}]\rangle \in h_{\underline{A_1}^{\mathfrak{S}}}} \langle \{s_{\alpha_1}\}, [\wp^L_{\gamma_{\underline{A_1}^{\mathfrak{S}}}}, \wp^U_{\gamma_{\underline{A_1}^{\mathfrak{S}}}}]\rangle$, and
$h_{\underline{A_2}^{\mathfrak{S}}} = \bigcup_{\langle s_{\alpha_2}, [\wp^L_{\gamma_{\underline{A_2}^{\mathfrak{S}}}}, \wp^U_{\gamma_{\underline{A_2}^{\mathfrak{S}}}}]\rangle \in h_{\underline{A_2}^{\mathfrak{S}}}} \langle \{s_{\alpha_2}\}, [\wp^L_{\gamma_{\underline{A_2}^{\mathfrak{S}}}}, \wp^U_{\gamma_{\underline{A_2}^{\mathfrak{S}}}}]\rangle$, some set and algebraic operations on the IPHFLVs, which are also IPHFLVs, can be described as follows (see, e.g., [20]):

$$h_{\underline{A_1}^{\mathfrak{S}}} \oplus h_{\underline{A_2}^{\mathfrak{S}}}$$
$$= \bigcup_{\langle s_{\alpha_1}, [\wp^L_{\gamma_{\underline{A_1}^{\mathfrak{S}}}}, \wp^U_{\gamma_{\underline{A_1}^{\mathfrak{S}}}}]\rangle \in h_{\underline{A_1}^{\mathfrak{S}}}, \langle s_{\alpha_2}, [\wp^L_{\gamma_{\underline{A_2}^{\mathfrak{S}}}}, \wp^U_{\gamma_{\underline{A_2}^{\mathfrak{S}}}}]\rangle \in h_{\underline{A_2}^{\mathfrak{S}}}} \langle \{s_{\alpha_1+\alpha_2}\},$$

$$
\begin{aligned}
&[\wp^L_{\gamma_{\underline{A}_1}\mathfrak{S}} + \wp^L_{\gamma_{\underline{A}_2}\mathfrak{S}} - \wp^L_{\gamma_{\underline{A}_1}\mathfrak{S}} \wp^L_{\gamma_{\underline{A}_2}\mathfrak{S}},\\
&\wp^U_{\gamma_{\underline{A}_1}\mathfrak{S}} + \wp^U_{\gamma_{\underline{A}_2}\mathfrak{S}} - \wp^U_{\gamma_{\underline{A}_1}\mathfrak{S}} \wp^U_{\gamma_{\underline{A}_2}\mathfrak{S}}]\rangle;
\end{aligned}
\tag{6.28}
$$

$$
\begin{aligned}
&h_{\underline{A}_1\mathfrak{S}} \otimes h_{\underline{A}_2\mathfrak{S}}\\
&= \bigcup_{\langle s_{\alpha_1}, [\wp^L_{\gamma_{\underline{A}_1}\mathfrak{S}}, \wp^U_{\gamma_{\underline{A}_1}\mathfrak{S}}]\rangle \in h_{\underline{A}_1\mathfrak{S}}, \langle s_{\alpha_2}, [\wp^L_{\gamma_{\underline{A}_2}\mathfrak{S}}, \wp^U_{\gamma_{\underline{A}_2}\mathfrak{S}}]\rangle \in h_{\underline{A}_2\mathfrak{S}}} \langle \{s_{\alpha_1 \times \alpha_2}\},\\
&[\wp^L_{\gamma_{\underline{A}_1}\mathfrak{S}} \wp^L_{\gamma_{\underline{A}_2}\mathfrak{S}},\\
&\wp^U_{\gamma_{\underline{A}_1}\mathfrak{S}} \wp^U_{\gamma_{\underline{A}_2}\mathfrak{S}}]\rangle;
\end{aligned}
\tag{6.29}
$$

$$
\begin{aligned}
&(h_{\underline{A}\mathfrak{S}})^{\lambda}\\
&= \bigcup_{\langle s_{\alpha}, [\wp^L_{\gamma_{\underline{A}}\mathfrak{S}}, \wp^U_{\gamma_{\underline{A}}\mathfrak{S}}]\rangle \in h_{\underline{A}\mathfrak{S}}} \langle \{s_{(\alpha)^{\lambda}}\}, [\wp^L_{\gamma_{\underline{A}}\mathfrak{S}}, \wp^U_{\gamma_{\underline{A}}\mathfrak{S}}]\rangle, \quad \lambda > 0;
\end{aligned}
\tag{6.30}
$$

$$
\begin{aligned}
&\lambda h_{\underline{A}\mathfrak{S}}\\
&= \bigcup_{\langle s_{\alpha}, [\wp^L_{\gamma_{\underline{A}}\mathfrak{S}}, \wp^U_{\gamma_{\underline{A}}\mathfrak{S}}]\rangle \in h_{\underline{A}\mathfrak{S}}} \langle \{s_{\lambda\alpha}\}, [\wp^L_{\gamma_{\underline{A}}\mathfrak{S}}, \wp^U_{\gamma_{\underline{A}}\mathfrak{S}}]\rangle, \quad \lambda > 0.
\end{aligned}
\tag{6.31}
$$

6.6 Probabilistic Neutrosophic Hesitant Fuzzy Set

Shao et al. [16] presented probabilistic neutrosophic hesitant fuzzy set, and they investigated the operation laws together with the averaging and geometric operators for probabilistic neutrosophic hesitant fuzzy sets. Furthermore, Shao et al. [17] developed Shao et al.'s [16] Choquet averaging and Choquet geometric operators for probabilistic neutrosophic hesitant fuzzy sets.

Definition 6.6 ([15]) Let X be a reference set. A probabilistic neutrosophic hesitant fuzzy set (PNHFS) $\overline{A}$ on X is defined in terms of three functions $u_{\overline{A}}(x)$, $w_{\overline{A}}(x)$, and $v_{\overline{A}}(x)$ as follows:

$$
\overline{A} = \{\langle x, \langle u_{\overline{A}}(x), w_{\overline{A}}(x), v_{\overline{A}}(x)\rangle\rangle \mid x \in X\},
\tag{6.32}
$$

where $u_{\overline{A}}(x), w_{\overline{A}}(x), v_{\overline{A}}(x) \in [0, 1]$ are, respectively, called truth, indeterminacy, and falsity membership degrees of the element $x \in X$ to $\overline{A}$ corresponding to their probabilistic information of factors in the forms of $u_{\overline{A}}(x) =$

$\bigcup_{\langle\gamma_{\overline{A}},\wp_{\gamma_{\overline{A}}}\rangle\in u_{\overline{A}}(x)}\{\langle\gamma_{\overline{A}},\wp_{\gamma_{\overline{A}}}\rangle\}$, $w_{\overline{A}}(x) = \bigcup_{\langle\delta_{\overline{A}},\wp_{\delta_{\overline{A}}}\rangle\in w_{\overline{A}}(x)}\{\langle\delta_{\overline{A}},\wp_{\delta_{\overline{A}}}\rangle\}$, and $\upsilon_{\overline{A}}(x) = \bigcup_{\langle\eta_{\overline{A}},\wp_{\eta_{\overline{A}}}\rangle\in \upsilon_{\overline{A}}(x)}\{\langle\eta_{\overline{A}},\wp_{\eta_{\overline{A}}}\rangle\}$.

We suppose that $\gamma_{\overline{A}}^{+} \in \bigcup_{x\in X}\max\{\gamma_{\overline{A}}(x)\}$, $\delta_{\overline{A}}^{+} \in \bigcup_{x\in X}\max\{\delta_{\overline{A}}(x)\}$ and $\eta_{\overline{A}}^{+} \in \bigcup_{x\in X}\max\{\eta_{\overline{A}}(x)\}$, and furthermore,

$$0 \le \gamma_{\overline{A}}, \delta_{\overline{A}}, \eta_{\overline{A}} \le 1, \quad 0 \le \gamma_{\overline{A}}^{+} + \delta_{\overline{A}}^{+} + \eta_{\overline{A}}^{+} \le 3.$$

Moreover, $\sum_{\langle\gamma_{\overline{A}},\wp_{\gamma_{\overline{A}}}\rangle\in u_{\overline{A}}(x)}\wp_{\gamma_{\overline{A}}} = \sum_{\langle\delta_{\overline{A}},\wp_{\delta_{\overline{A}}}\rangle\in w_{\overline{A}}(x)}\wp_{\delta_{\overline{A}}} = \sum_{\langle\eta_{\overline{A}},\wp_{\eta_{\overline{A}}}\rangle\in \upsilon_{\overline{A}}(x)}\wp_{\eta_{\overline{A}}} = 1$ for any $0 \le \wp_{\gamma_{\overline{A}}}, \wp_{\delta_{\overline{A}}}, \wp_{\eta_{\overline{A}}} \le 1$.

Remark 6.7 Throughout this book, the set of all PNHFSs on the reference set X is denoted by $\mathbb{PNHFS}(X)$.

Example 6.6 Let $X = \{x_1, x_2\}$ be the reference set, $h_{\overline{A}}(x_1) = \{\langle\{\langle 0.2, 0.7\rangle, \langle 0.3, 0.3\rangle\}, \{\langle 0.4, 0.5\rangle, \langle 0.3, 0.5\rangle\}, \{\langle 0.3, 1.0\rangle\}\rangle\}$ and $h_{\overline{A}}(x_2) = \{\langle\{\langle 0.2, 1.0\rangle\}, \{\langle 0.3, 0.5\rangle, \langle 0.4, 0.5\rangle\}, \{\langle 0.3, 0.2\rangle, \langle 0.6, 0.8\rangle\}\rangle\}$ be the PNHFEs of x_i $(i = 1, 2)$ to a set $\overline{A}$, respectively. Then $\overline{A}$ can be considered as an PNHFS, i.e.,

$$\begin{aligned}\overline{A} = \{&\langle x_1, \langle\{\langle 0.2, 0.7\rangle, \langle 0.3, 0.3\rangle\}, \{\langle 0.4, 0.5\rangle, \langle 0.3, 0.5\rangle\}, \{\langle 0.3, 1.0\rangle\}\rangle\rangle,\\ &\langle x_2, \langle\{\langle 0.2, 1.0\rangle\}, \{\langle 0.3, 0.5\rangle, \langle 0.4, 0.5\rangle\}, \{\langle 0.3, 0.2\rangle, \langle 0.6, 0.8\rangle\}\rangle\rangle\}.\end{aligned}$$

Given three PNHFEs represented by
$h_{\overline{A}} = \langle u_{\overline{A}}, w_{\overline{A}}, \upsilon_{\overline{A}}\rangle = \langle\bigcup_{\langle\gamma_{\overline{A}},\wp_{\gamma_{\overline{A}}}\rangle\in u_{\overline{A}}}\{\langle\gamma_{\overline{A}},\wp_{\gamma_{\overline{A}}}\rangle\}, \bigcup_{\langle\delta_{\overline{A}},\wp_{\delta_{\overline{A}}}\rangle\in w_{\overline{A}}}\{\langle\delta_{\overline{A}},\wp_{\delta_{\overline{A}}}\rangle\}, \bigcup_{\langle\eta_{\overline{A}},\wp_{\eta_{\overline{A}}}\rangle\in \upsilon_{\overline{A}}}\{\langle\eta_{\overline{A}},\wp_{\eta_{\overline{A}}}\rangle\}\rangle$,
$h_{\overline{A}_1} = \langle u_{\overline{A}_1}, w_{\overline{A}_1}, \upsilon_{\overline{A}_1}\rangle = \langle\bigcup_{\langle\gamma_{\overline{A}_1},\wp_{\gamma_{\overline{A}_1}}\rangle\in u_{\overline{A}_1}}\{\langle\gamma_{\overline{A}_1},\wp_{\gamma_{\overline{A}_1}}\rangle\}, \bigcup_{\langle\delta_{\overline{A}_1},\wp_{\delta_{\overline{A}_1}}\rangle\in w_{\overline{A}_1}}\{\langle\delta_{\overline{A}_1},\wp_{\delta_{\overline{A}_1}}\rangle\}, \bigcup_{\langle\eta_{\overline{A}_1},\wp_{\eta_{\overline{A}_1}}\rangle\in \upsilon_{\overline{A}_1}}\{\langle\eta_{\overline{A}_1},\wp_{\eta_{\overline{A}_1}}\rangle\}\rangle$ and
$h_{\overline{A}_2} = \langle u_{\overline{A}_2}, w_{\overline{A}_2}, \upsilon_{\overline{A}_2}\rangle = \langle\bigcup_{\langle\gamma_{\overline{A}_2},\wp_{\gamma_{\overline{A}_2}}\rangle\in u_{\overline{A}_2}}\{\langle\gamma_{\overline{A}_2},\wp_{\gamma_{\overline{A}_2}}\rangle\}, \bigcup_{\langle\delta_{\overline{A}_2},\wp_{\delta_{\overline{A}_2}}\rangle\in w_{\overline{A}_2}}\{\langle\delta_{\overline{A}_2},\wp_{\delta_{\overline{A}_2}}\rangle\}, \bigcup_{\langle\eta_{\overline{A}_2},\wp_{\eta_{\overline{A}_2}}\rangle\in \upsilon_{\overline{A}_2}}\{\langle\eta_{\overline{A}_2},\wp_{\eta_{\overline{A}_2}}\rangle\}\rangle$, a number of set and algebraic operations on the PNHFEs, which are also PNHFEs, can be described as follows (see, e.g., [18]):

$$\begin{aligned}h_{\overline{A}_1}\cup h_{\overline{A}_2} = \Bigg\langle &\bigcup_{\langle\gamma_{\overline{A}1},\wp_{\gamma_{\overline{A}1}}\rangle\in u_{\overline{A}_1},\langle\gamma_{\overline{A}2},\wp_{\gamma_{\overline{A}2}}\rangle\in u_{\overline{A}_2}}\{\langle\max\{\gamma_{\overline{A}1},\gamma_{\overline{A}2}\},\wp_{\gamma_{\overline{A}1}}\wp_{\gamma_{\overline{A}2}}\rangle\},\\ &\bigcup_{\langle\delta_{\overline{A}1},\wp_{\delta_{\overline{A}1}}\rangle\in w_{\overline{A}_1},\langle\delta_{\overline{A}2},\wp_{\delta_{\overline{A}2}}\rangle\in w_{\overline{A}_2}}\{\langle\min\{\delta_{\overline{A}1},\delta_{\overline{A}2}\},\wp_{\delta_{\overline{A}1}}\wp_{\delta_{\overline{A}2}}\rangle\},\\ &\bigcup_{\langle\eta_{\overline{A}1},\wp_{\eta_{\overline{A}1}}\rangle\in \upsilon_{\overline{A}_1},\langle\eta_{\overline{A}2},\wp_{\eta_{\overline{A}2}}\rangle\in \upsilon_{\overline{A}_2}}\{\langle\min\{\eta_{\overline{A}1},\eta_{\overline{A}2}\},\wp_{\eta_{\overline{A}1}}\wp_{\eta_{\overline{A}2}}\rangle\}\Bigg\rangle;\end{aligned} \tag{6.33}$$

$$h_{\overline{A}_1} \cap h_{\overline{A}_2} = \left\langle \bigcup_{\langle \gamma_{\overline{A}1}, \wp_{\gamma_{\overline{A}1}} \rangle \in u_{\overline{A}_1}, \langle \gamma_{\overline{A}2}, \wp_{\gamma_{\overline{A}2}} \rangle \in u_{\overline{A}_2}} \{\langle \min\{\gamma_{\overline{A}1}, \gamma_{\overline{A}2}\}, \wp_{\gamma_{\overline{A}1}} \wp_{\gamma_{\overline{A}2}} \rangle\}, \right.$$

$$\bigcup_{\langle \delta_{\overline{A}1}, \wp_{\delta_{\overline{A}1}} \rangle \in w_{\overline{A}_1}, \langle \delta_{\overline{A}2}, \wp_{\delta_{\overline{A}2}} \rangle \in w_{\overline{A}_2}} \{\langle \max\{\delta_{\overline{A}1}, \delta_{\overline{A}2}\}, \wp_{\delta_{\overline{A}1}} \wp_{\delta_{\overline{A}2}} \rangle\},$$

$$\left. \bigcup_{\langle \eta_{\overline{A}1}, \wp_{\eta_{\overline{A}1}} \rangle \in v_{\overline{A}_1}, \langle \eta_{\overline{A}2}, \wp_{\eta_{\overline{A}2}} \rangle \in v_{\overline{A}_2}} \{\langle \max\{\eta_{\overline{A}1}, \eta_{\overline{A}2}\}, \wp_{\eta_{\overline{A}1}} \wp_{\eta_{\overline{A}2}} \rangle\} \right\rangle; \tag{6.34}$$

$$h_{\overline{A}_1} \oplus h_{\overline{A}_2} = \left\langle \bigcup_{\langle \gamma_{\overline{A}1}, \wp_{\gamma_{\overline{A}1}} \rangle \in u_{\overline{A}_1}, \langle \gamma_{\overline{A}2}, \wp_{\gamma_{\overline{A}2}} \rangle \in u_{\overline{A}_2}} \{\langle \gamma_{\overline{A}1} + \gamma_{\overline{A}2} - \gamma_{\overline{A}1} \times \gamma_{\overline{A}2}, \wp_{\gamma_{\overline{A}1}} \wp_{\gamma_{\overline{A}2}} \rangle\}, \right.$$

$$\bigcup_{\langle \delta_{\overline{A}1}, \wp_{\delta_{\overline{A}1}} \rangle \in w_{\overline{A}_1}, \langle \delta_{\overline{A}2}, \wp_{\delta_{\overline{A}2}} \rangle \in w_{\overline{A}_2}} \{\langle \delta_{\overline{A}1} \times \delta_{\overline{A}2}, \wp_{\delta_{\overline{A}1}} \wp_{\delta_{\overline{A}2}} \rangle\},$$

$$\left. \bigcup_{\langle \eta_{\overline{A}1}, \wp_{\eta_{\overline{A}1}} \rangle \in v_{\overline{A}_1}, \langle \eta_{\overline{A}2}, \wp_{\eta_{\overline{A}2}} \rangle \in v_{\overline{A}_2}} \{\langle \eta_{\overline{A}1} \times \eta_{\overline{A}2}\}, \wp_{\eta_{\overline{A}1}} \wp_{\eta_{\overline{A}2}} \rangle\} \right\rangle; \tag{6.35}$$

$$h_{\overline{A}_1} \otimes h_{\overline{A}_2} = \left\langle \bigcup_{\langle \gamma_{\overline{A}1}, \wp_{\gamma_{\overline{A}1}} \rangle \in u_{\overline{A}_1}, \langle \gamma_{\overline{A}2}, \wp_{\gamma_{\overline{A}2}} \rangle \in u_{\overline{A}_2}} \{\langle \gamma_{\overline{A}1} \times \gamma_{\overline{A}2}, \wp_{\gamma_{\overline{A}1}} \wp_{\gamma_{\overline{A}2}} \rangle\}, \right.$$

$$\bigcup_{\langle \delta_{\overline{A}1}, \wp_{\delta_{\overline{A}1}} \rangle \in w_{\overline{A}_1}, \langle \delta_{\overline{A}2}, \wp_{\delta_{\overline{A}2}} \rangle \in w_{\overline{A}_2}} \{\langle \delta_{\overline{A}1} + \delta_{\overline{A}2} - \delta_{\overline{A}1} \times \delta_{\overline{A}2}, \wp_{\delta_{\overline{A}1}} \wp_{\delta_{\overline{A}2}} \rangle\},$$

$$\left. \bigcup_{\langle \eta_{\overline{A}1}, \wp_{\eta_{\overline{A}1}} \rangle \in v_{\overline{A}_1}, \langle \eta_{\overline{A}2}, \wp_{\eta_{\overline{A}2}} \rangle \in v_{\overline{A}_2}} \{\langle \eta_{\overline{A}1} + \eta_{\overline{A}2} - \eta_{\overline{A}1} \times \eta_{\overline{A}2}\}, \wp_{\eta_{\overline{A}1}} \wp_{\eta_{\overline{A}2}} \rangle\} \right\rangle; \tag{6.36}$$

$$\lambda h_{\overline{A}} = \left\langle \bigcup_{\langle \gamma_{\overline{A}}, \wp_{\gamma_{\overline{A}}} \rangle \in u_{\overline{A}}} \{\langle 1 - (1 - \gamma_{\overline{A}})^{\lambda}, \wp_{\gamma_{\overline{A}}} \rangle\}, \right.$$

$$\bigcup_{\langle \delta_{\overline{A}}, \wp_{\delta_{\overline{A}}} \rangle \in w_{\overline{A}}} \{\langle (\delta_{\overline{A}})^{\lambda}, \wp_{\delta_{\overline{A}}} \rangle\},$$

$$\left. \bigcup_{\langle \eta_{\overline{A}}, \wp_{\eta_{\overline{A}}} \rangle \in v_{\overline{A}}} \{\langle (\eta_{\overline{A}})^{\lambda}\}, \wp_{\eta_{\overline{A}}} \rangle\} \right\rangle, \lambda > 0; \tag{6.37}$$

$$h^{\lambda}_{\overline{A}} = \left\langle \bigcup_{\langle \gamma_{\overline{A}}, \wp_{\gamma_{\overline{A}}} \rangle \in u_{\overline{A}}} \{\langle (\gamma_{\overline{A}})^{\lambda}, \wp_{\gamma_{\overline{A}}} \rangle\}, \right.$$

$$\bigcup_{\langle \delta_{\overline{A}}, \wp_{\delta_{\overline{A}}} \rangle \in w_{\overline{A}}} \{\langle 1-(1-\delta_{\overline{A}})^{\lambda}, \wp_{\delta_{\overline{A}}}\rangle\},$$

$$\left.\bigcup_{\langle \eta_{\overline{A}}, \wp_{\eta_{\overline{A}}} \rangle \in v_{\overline{A}}} \{\langle 1-(1-\eta_{\overline{A}})^{\lambda}\}, \wp_{\eta_{\overline{A}}}\rangle\}\right\rangle, \lambda > 0. \tag{6.38}$$

6.7 Pythagorean Probabilistic Hesitant Fuzzy Set

Batool et al. [1] introduced the concept of Pythagorean probabilistic hesitant fuzzy set with the constraint that the square sum of positive and negative hesitant membership degrees is less than or equal to one. Batool et al. [2] established a multiple criteria decision making technique on the basis of EDAS approach under Pythagorean probabilistic hesitant fuzzy information, and further, they developed an algorithm for addressing the uncertainty in the selection of drugs in EmDM issues in correspondence with the clinical analysis.

Definition 6.7 ([1]) Let X be a reference set. A Pythagorean probabilistic hesitant fuzzy set (PPHFS) $\acute{A}$ on X is defined in terms of two functions $u_{\acute{A}}(x)$ and $v_{\acute{A}}(x)$ as follows:

$$\acute{A} = \{\langle x, u_{\acute{A}}(x), v_{\acute{A}}(x)\rangle \mid x \in X\}, \tag{6.39}$$

where $u_{\acute{A}}(x)$ and $v_{\acute{A}}(x)$ are some different probabilistic values in [0, 1] representing the possible probabilistic membership degrees and probabilistic non-membership degrees of the element $x \in X$ to the set $\acute{A}$, respectively.

Here, for all $x \in X$, if we take $u_{\acute{A}}(x) = \bigcup_{\langle \gamma_{\acute{A}}, \wp_{\gamma_{\acute{A}}}\rangle \in u_{\acute{A}}(x)} \{\langle \gamma_{\acute{A}}, \wp_{\gamma_{\acute{A}}}\rangle\}$, $v_{\acute{A}}(x) = \bigcup_{\langle \eta_{\acute{A}}, \wp_{\eta_{\acute{A}}}\rangle \in v_{\acute{A}}(x)} \{\langle \eta_{\acute{A}}, \wp_{\eta_{\acute{A}}}\rangle\}$, $\gamma_{\acute{A}}^{+} \in \bigcup_{x \in X} \max\{\gamma_{\acute{A}}(x)\}$ and $\eta_{\acute{A}}^{+} \in \bigcup_{x \in X} \max\{\eta_{\acute{A}}(x)\}$, then we find that

$$0 \leq \gamma_{\acute{A}}, \eta_{\acute{A}} \leq 1, \quad 0 \leq (\gamma_{\acute{A}}^{+})^2 + (\eta_{\acute{A}}^{+})^2 \leq 1.$$

Moreover, $\sum_{\langle \gamma_{\acute{A}}, \wp_{\gamma_{\acute{A}}}\rangle \in u_{\acute{A}}(x)} \wp_{\gamma_{\acute{A}}} = \sum_{\langle \eta_{\acute{A}}, \wp_{\eta_{\acute{A}}}\rangle \in v_{\acute{A}}(x)} \wp_{\eta_{\acute{A}}} = 1$ for any $0 \leq \wp_{\gamma_{\acute{A}}}, \wp_{\eta_{\acute{A}}} \leq 1$.

For the sake of simplicity, $h_{\acute{A}}(x) = (u_{\acute{A}}(x), v_{\acute{A}}(x))$ is called the Pythagorean probabilistic hesitant fuzzy element (PPHFE).

Remark 6.8 Throughout this book, the set of all PPHFSs on the reference set X is denoted by $\mathbb{PPHFS}(X)$.

Example 6.7 Let $X = \{x_1, x_2\}$ be the reference set,
$h_{\acute{A}}(x_1) = (\bigcup_{\langle\gamma_{\acute{A}},\wp\gamma_{\acute{A}}\rangle\in u_{\acute{A}}(x_1)}\{\langle\gamma_{\acute{A}},\wp\gamma_{\acute{A}}\rangle\}, \bigcup_{\langle\eta_{\acute{A}},\wp\eta_{\acute{A}}\rangle\in v_{\acute{A}}(x_1)}\{\langle\eta_{\acute{A}},\wp\eta_{\acute{A}}\rangle\}$
$= (\{\langle 0.2, 0.7\rangle, \langle 0.5, 0.3\rangle\}, \{\langle 0.7, 1\rangle\})$ and
$h_{\acute{A}}(x_2) = (\bigcup_{\langle\gamma_{\acute{A}},\wp\gamma_{\acute{A}}\rangle\in u_{\acute{A}}(x_2)}\{\langle\gamma_{\acute{A}},\wp\gamma_{\acute{A}}\rangle\}, \bigcup_{\langle\eta_{\acute{A}},\wp\eta_{\acute{A}}\rangle\in v_{\acute{A}}(x_2)}\{\langle\eta_{\acute{A}},\wp\eta_{\acute{A}}\rangle\}$
$= (\{\langle 0.3, 0.2\rangle, \langle 0.5, 0.8\rangle\}, \{\langle 0.2, 1\rangle\})$ be the PDHFEs of x_i $(i = 1, 2)$ in the set $\acute{A}$, respectively. In this case,

$$(\gamma_{\acute{A}}{}^+)(x_1) = 0.5,\ (\eta_{\acute{A}}{}^+)(x_1) = 0.7;$$
$$(\gamma_{\acute{A}}{}^+)(x_2) = 0.5,\ (\eta_{\acute{A}}{}^+)(x_2) = 0.2;$$

result in

$$(\gamma_{\acute{A}}{}^+)^2(x_1) + (\eta_{\acute{A}}{}^+)^2(x_1) = 0.5^2 + 0.7^2 \leq 1;$$
$$(\gamma_{\acute{A}}{}^+)^2(x_2) + (\eta_{\acute{A}}{}^+)^2(x_2) = 0.5^2 + 0.2^2 \leq 1.$$

Thus, $h_{\acute{A}}(x_1)$ and $h_{\acute{A}}(x_2)$ are two PPHFEs, and then $\acute{A}$ can be considered as a PPHFS, i.e.,

$$\acute{A} = \{\langle x_1, (\{\langle 0.2, 0.7\rangle, \langle 0.5, 0.3\rangle\}, \{\langle 0.7, 1\rangle\})\rangle,$$
$$\langle x_2, (\{\langle 0.3, 0.2\rangle, \langle 0.5, 0.8\rangle\}, \{\langle 0.2, 1\rangle\})\rangle\}.$$

For the PPHFEs
$h_{\acute{A}} = (u_{\acute{A}}, v_{\acute{A}}) = (\bigcup_{\langle\gamma_{\acute{A}},\wp\gamma_{\acute{A}}\rangle\in u_{\acute{A}}}\{\langle\gamma_{\acute{A}},\wp\gamma_{\acute{A}}\rangle\}, \bigcup_{\langle\eta_{\acute{A}},\wp\eta_{\acute{A}}\rangle\in v_{\acute{A}}}\{\langle\eta_{\acute{A}},\wp\eta_{\acute{A}}\rangle\})$,
$h_{\acute{A}_1} = (u_{\acute{A}_1}, v_{\acute{A}_1}) = (\bigcup_{\langle\gamma_{\acute{A}_1},\wp\gamma_{\acute{A}_1}\rangle\in u_{\acute{A}_1}}\{\langle\gamma_{\acute{A}_1},\wp\gamma_{\acute{A}_1}\rangle\}, \bigcup_{\langle\eta_{\acute{A}_1},\wp\eta_{\acute{A}_1}\rangle\in v_{\acute{A}_1}}\{\langle\eta_{\acute{A}_1},\wp\eta_{\acute{A}_1}\rangle\})$
and
$h_{\acute{A}_2} = (u_{\acute{A}_2}, v_{\acute{A}_2}) = (\bigcup_{\langle\gamma_{\acute{A}_2},\wp\gamma_{\acute{A}_2}\rangle\in u_{\acute{A}_2}}\{\langle\gamma_{\acute{A}_2},\wp\gamma_{\acute{A}_2}\rangle\}, \bigcup_{\langle\eta_{\acute{A}_2},\wp\eta_{\acute{A}_2}\rangle\in v_{\acute{A}_2}}\{\langle\eta_{\acute{A}_2},\wp\eta_{\acute{A}_2}\rangle\})$
the following operations are defined (see [1]):

$$h_{\acute{A}_1} \oplus h_{\acute{A}_2} = \left(\bigcup_{\langle\gamma_{\acute{A}_1},\wp\gamma_{\acute{A}_1}\rangle\in u_{\acute{A}_1},\langle\gamma_{\acute{A}_2},\wp\gamma_{\acute{A}_2}\rangle\in u_{\acute{A}_2}} \{\langle\sqrt[2]{\gamma_{\acute{A}_1}{}^2+\gamma_{\acute{A}_2}{}^2-\gamma_{\acute{A}_1}{}^2\times\gamma_{\acute{A}_2}{}^2}, \wp\gamma_{\acute{A}_1}\times\wp\gamma_{\acute{A}_2}\rangle\}, \bigcup_{\langle\eta_{\acute{A}_1},\wp\eta_{\acute{A}_1}\rangle\in v_{\acute{A}_1},\langle\eta_{\acute{A}_2},\wp\eta_{\acute{A}_2}\rangle\in v_{\acute{A}_2}} \{\langle\eta_{\acute{A}_1}\times\eta_{\acute{A}_2}, \wp\eta_{\acute{A}_1}\times\wp\eta_{\acute{A}_2}\rangle\} \right); \tag{6.40}$$

$$h_{\acute{A}_1} \otimes h_{\acute{A}_2} = \left(\bigcup_{\langle \gamma_{\acute{A}_1}, \wp\gamma_{\acute{A}_1} \rangle \in u_{\acute{A}_1}, \langle \gamma_{\acute{A}_2}, \wp\gamma_{\acute{A}_2} \rangle \in u_{\acute{A}_2}} \{\langle \gamma_{\acute{A}_1} \times \gamma_{\acute{A}_2}, \wp\gamma_{\acute{A}_1} \times \wp\gamma_{\acute{A}_2} \rangle\}, \bigcup_{\langle \eta_{\acute{A}_1}, \wp\eta_{\acute{A}_1} \rangle \in v_{\acute{A}_1}, \langle \eta_{\acute{A}_2}, \wp\eta_{\acute{A}_2} \rangle \in v_{\acute{A}_2}} \{\langle \sqrt[2]{\eta_{\acute{A}_1}{}^2 + \eta_{\acute{A}_2}{}^2 - \eta_{\acute{A}_1}{}^2 \times \eta_{\acute{A}_2}{}^2}, \wp\eta_{\acute{A}_1} \times \wp\eta_{\acute{A}_2} \rangle\} \right); \tag{6.41}$$

$$\lambda h_{\acute{A}} = \left(\bigcup_{\langle \gamma_{\acute{A}}, \wp\gamma_{\acute{A}} \rangle \in u_{\acute{A}}} \{\langle \sqrt[2]{1 - (1 - \gamma_{\acute{A}}{}^2)^{\lambda}}, \wp\gamma_{\acute{A}} \rangle\}, \bigcup_{\langle \eta_{\acute{A}}, \wp\eta_{\acute{A}} \rangle \in v_{\acute{A}}} \{\langle (\eta_{\acute{A}})^{\lambda}, \wp\eta_{\acute{A}} \rangle\} \right); \tag{6.42}$$

$$h_{\acute{A}}{}^{\lambda} = \left(\bigcup_{\langle \gamma_{\acute{A}}, \wp\gamma_{\acute{A}} \rangle \in u_{\acute{A}}} \{\langle (\gamma_{\acute{A}})^{\lambda}, \wp\gamma_{\acute{A}} \rangle\}, \bigcup_{\langle \eta_{\acute{A}}, \wp\eta_{\acute{A}} \rangle \in v_{\acute{A}}} \{\langle \sqrt[2]{1 - (1 - \eta_{\acute{A}}{}^2)^{\lambda}}, \wp\eta_{\acute{A}} \rangle\} \right). \tag{6.43}$$

6.8 q-Rung Orthopair Probabilistic Hesitant Fuzzy Set

Li et al. [11] introduced the concept of q-rung probabilistic dual hesitant fuzzy set by the use of capturing the probability of each element in q-rung dual hesitant fuzzy sets. Q-rung probabilistic dual hesitant fuzzy set is indeed q-rung probabilistic hesitant fuzzy set. The most prominent character of q-rung probabilistic dual hesitant fuzzy set is that it allows experts to have a flexible manner for evaluating criterion values in a complicated and realistic multiple criteria decision making situation. In order to more effectively exploit the opportunities for applying q-rung probabilistic dual hesitant fuzzy sets, Li et al. [11] proposed some basic operational rules, comparison technique and distance measure of q-rung probabilistic dual hesitant fuzzy sets.

Definition 6.8 ([11]) Let X be a reference set. A q-rung orthopair probabilistic hesitant fuzzy set (q-ROPHFS) $\grave{A}$ on X is defined in terms of two functions $u_{\grave{A}}(x)$ and $v_{\grave{A}}(x)$ as follows:

$$\grave{A} = \{\langle x, u_{\grave{A}}(x), v_{\grave{A}}(x)\rangle \mid x \in X\}, \tag{6.44}$$

where $u_{\grave{A}}(x)$ and $v_{\grave{A}}(x)$ are some different probabilistic values in [0, 1] representing the possible probabilistic membership degrees and probabilistic non-membership degrees of the element $x \in X$ to the set $\grave{A}$, respectively.

Here, for all $x \in X$, if we consider $u_{\grave{A}}(x) = \bigcup_{\langle \gamma_{\grave{A}}, \wp_{\gamma_{\grave{A}}}\rangle \in u_{\grave{A}}(x)} \{\langle \gamma_{\grave{A}}, \wp_{\gamma_{\grave{A}}}\rangle\}$, $v_{\grave{A}}(x) = \bigcup_{\langle \eta_{\grave{A}}, \wp_{\eta_{\grave{A}}}\rangle \in v_{\grave{A}}(x)} \{\langle \eta_{\grave{A}}, \wp_{\eta_{\grave{A}}}\rangle\}$, ${\gamma_{\grave{A}}}^+ \in \bigcup_{x \in X} \max\{\gamma_{\grave{A}}(x)\}$ and ${\eta_{\grave{A}}}^+ \in \bigcup_{x \in X} \max\{\eta_{\grave{A}}(x)\}$, then we conclude that

$$0 \le \gamma_{\grave{A}}, \eta_{\grave{A}} \le 1, \quad 0 \le ({\gamma_{\grave{A}}}^+)^q + ({\eta_{\grave{A}}}^+)^q \le 1, \quad q \ge 1.$$

Moreover, $\sum_{\langle \gamma_{\grave{A}}, \wp_{\gamma_{\grave{A}}}\rangle \in u_{\grave{A}}(x)} \wp_{\gamma_{\grave{A}}} = \sum_{\langle \eta_{\grave{A}}, \wp_{\eta_{\grave{A}}}\rangle \in v_{\grave{A}}(x)} \wp_{\eta_{\grave{A}}} = 1$ for any $0 \le \wp_{\gamma_{\grave{A}}}, \wp_{\eta_{\grave{A}}} \le 1$.

For the sake of simplicity, $h_{\grave{A}}(x) = (u_{\grave{A}}(x), v_{\grave{A}}(x))$ is called the q-rung orthopair probabilistic hesitant fuzzy element (q-ROPHFE).

Remark 6.9 Throughout this book, the set of all q-ROPHFSs on the reference set X is denoted by $q - \mathbb{ROPHFS}(X)$.

Example 6.8 Let $X = \{x_1, x_2\}$ be the reference set, $h_{\grave{A}}(x_1) = (\bigcup_{\langle \gamma_{\grave{A}}, \wp_{\gamma_{\grave{A}}}\rangle \in u_{\grave{A}}(x_1)} \{\langle \gamma_{\grave{A}}, \wp_{\gamma_{\grave{A}}}\rangle\}, \bigcup_{\langle \eta_{\grave{A}}, \wp_{\eta_{\grave{A}}}\rangle \in v_{\grave{A}}(x_1)} \{\langle \eta_{\grave{A}}, \wp_{\eta_{\grave{A}}}\rangle\}$ $= (\{\langle 0.2, 0.7\rangle, \langle 0.8, 0.3\rangle\}, \{\langle 0.7, 1\rangle\})$ and $h_{\grave{A}}(x_2) = (\bigcup_{\langle \gamma_{\grave{A}}, \wp_{\gamma_{\grave{A}}}\rangle \in u_{\grave{A}}(x_2)} \{\langle \gamma_{\grave{A}}, \wp_{\gamma_{\grave{A}}}\rangle\}, \bigcup_{\langle \eta_{\grave{A}}, \wp_{\eta_{\grave{A}}}\rangle \in v_{\grave{A}}(x_2)} \{\langle \eta_{\grave{A}}, \wp_{\eta_{\grave{A}}}\rangle\}$ $= (\{\langle 0.3, 0.2\rangle, \langle 0.5, 0.8\rangle\}, \{\langle 0.2, 1\rangle\})$ be the q-ROPHFEs of x_i $(i = 1, 2)$ in the set $\grave{A}$, respectively. In this case,

$$({\gamma_{\grave{A}}}^+)(x_1) = 0.8, \ ({\eta_{\grave{A}}}^+)(x_1) = 0.7;$$

$$({\gamma_{\grave{A}}}^+)(x_2) = 0.5, \ ({\eta_{\grave{A}}}^+)(x_2) = 0.2;$$

result in

$$({\gamma_{\grave{A}}}^+)^q(x_1) + ({\eta_{\grave{A}}}^+)^q(x_1) = 0.8^q + 0.7^q \le 1;$$

$$({\gamma_{\grave{A}}}^+)^q(x_2) + ({\eta_{\grave{A}}}^+)^q(x_2) = 0.5^q + 0.2^q \le 1,$$

which hold true for any $q \ge 3$. Thus, $h_{\grave{A}}(x_1)$ and $h_{\grave{A}}(x_2)$ are two 3-ROPHFEs, and then $\grave{A}$ can be considered as a q-ROPHFS, i.e.,

$$\grave{A} = \{\langle x_1, (\{\langle 0.2, 0.7\rangle, \langle 0.8, 0.3\rangle\}, \{\langle 0.7, 1\rangle\})\rangle,$$
$$\langle x_2, (\{\langle 0.3, 0.2\rangle, \langle 0.5, 0.8\rangle\}, \{\langle 0.2, 1\rangle\})\rangle\}.$$

For the q-ROPHFEs
$h_{\grave{A}} = (u_{\grave{A}}, v_{\grave{A}}) = (\bigcup_{\langle\gamma_{\grave{A}},\wp_{\gamma_{\grave{A}}}\rangle\in u_{\grave{A}}}\{\langle\gamma_{\grave{A}}, \wp_{\gamma_{\grave{A}}}\rangle\}, \bigcup_{\langle\eta_{\grave{A}},\wp_{\eta_{\grave{A}}}\rangle\in v_{\grave{A}}}\{\langle\eta_{\grave{A}}, \wp_{\eta_{\grave{A}}}\rangle\})$,
$h_{\grave{A}_1} = (u_{\grave{A}_1}, v_{\grave{A}_1}) = (\bigcup_{\langle\gamma_{\grave{A}_1},\wp_{\gamma_{\grave{A}_1}}\rangle\in u_{\grave{A}_1}}\{\langle\gamma_{\grave{A}_1}, \wp_{\gamma_{\grave{A}_1}}\rangle\}, \bigcup_{\langle\eta_{\grave{A}_1},\wp_{\eta_{\grave{A}_1}}\rangle\in v_{\grave{A}_1}}\{\langle\eta_{\grave{A}_1}, \wp_{\eta_{\grave{A}_1}}\rangle\})$
and
$h_{\grave{A}_2} = (u_{\grave{A}_2}, v_{\grave{A}_2}) = (\bigcup_{\langle\gamma_{\grave{A}_2},\wp_{\gamma_{\grave{A}_2}}\rangle\in u_{\grave{A}_2}}\{\langle\gamma_{\grave{A}_2}, \wp_{\gamma_{\grave{A}_2}}\rangle\}, \bigcup_{\langle\eta_{\grave{A}_2},\wp_{\eta_{\grave{A}_2}}\rangle\in v_{\grave{A}_2}}\{\langle\eta_{\grave{A}_2}, \wp_{\eta_{\grave{A}_2}}\rangle\})$
the following operations are defined (see [1]):

$$h_{\grave{A}_1} \oplus h_{\grave{A}_2} = \left(\bigcup_{\langle\gamma_{\grave{A}_1},\wp_{\gamma_{\grave{A}_1}}\rangle\in u_{\grave{A}_1},\langle\gamma_{\grave{A}_2},\wp_{\gamma_{\grave{A}_2}}\rangle\in u_{\grave{A}_2}} \left\{\langle \sqrt[q]{{\gamma_{\grave{A}_1}}^q + {\gamma_{\grave{A}_2}}^q - {\gamma_{\grave{A}_1}}^q \times {\gamma_{\grave{A}_2}}^q}, \wp_{\gamma_{\grave{A}_1}} \times \wp_{\gamma_{\grave{A}_2}}\rangle\right\}, \bigcup_{\langle\eta_{\grave{A}_1},\wp_{\eta_{\grave{A}_1}}\rangle\in v_{\grave{A}_1},\langle\eta_{\grave{A}_2},\wp_{\eta_{\grave{A}_2}}\rangle\in v_{\grave{A}_2}} \{\langle\eta_{\grave{A}_1} \times \eta_{\grave{A}_2}, \wp_{\eta_{\grave{A}_1}} \times \wp_{\eta_{\grave{A}_2}}\rangle\} \right); \tag{6.45}$$

$$h_{\grave{A}_1} \otimes h_{\grave{A}_2} = \left(\bigcup_{\langle\gamma_{\grave{A}_1},\wp_{\gamma_{\grave{A}_1}}\rangle\in u_{\grave{A}_1},\langle\gamma_{\grave{A}_2},\wp_{\gamma_{\grave{A}_2}}\rangle\in u_{\grave{A}_2}} \{\langle\gamma_{\grave{A}_1} \times \gamma_{\grave{A}_2}, \wp_{\gamma_{\grave{A}_1}} \times \wp_{\gamma_{\grave{A}_2}}\rangle\}, \bigcup_{\langle\eta_{\grave{A}_1},\wp_{\eta_{\grave{A}_1}}\rangle\in v_{\grave{A}_1},\langle\eta_{\grave{A}_2},\wp_{\eta_{\grave{A}_2}}\rangle\in v_{\grave{A}_2}} \left\{\langle \sqrt[q]{{\eta_{\grave{A}_1}}^q + {\eta_{\grave{A}_2}}^q - {\eta_{\grave{A}_1}}^q \times {\eta_{\grave{A}_2}}^q}, \wp_{\eta_{\grave{A}_1}} \times \wp_{\eta_{\grave{A}_2}}\rangle\right\} \right); \tag{6.46}$$

$$\lambda h_{\grave{A}} = \left(\bigcup_{\langle\gamma_{\grave{A}},\wp_{\gamma_{\grave{A}}}\rangle\in u_{\grave{A}}} \{\langle \sqrt[q]{1 - (1 - {\gamma_{\grave{A}}}^q)^\lambda}, \wp_{\gamma_{\grave{A}}}\rangle\}, \bigcup_{\langle\eta_{\grave{A}},\wp_{\eta_{\grave{A}}}\rangle\in v_{\grave{A}}} \{\langle(\eta_{\grave{A}})^\lambda, \wp_{\eta_{\grave{A}}}\rangle\} \right); \tag{6.47}$$

$$h_{\grave{A}}{}^{\lambda} = \left(\bigcup_{\langle \gamma_{\grave{A}}, \wp_{\gamma_{\grave{A}}} \rangle \in u_{\grave{A}}} \{\langle (\gamma_{\grave{A}})^{\lambda}, \wp_{\gamma_{\grave{A}}} \rangle\}, \bigcup_{\langle \eta_{\grave{A}}, \wp_{\eta_{\grave{A}}} \rangle \in v_{\grave{A}}} \{\langle \sqrt[q]{1 - (1 - \eta_{\grave{A}}{}^{q})^{\lambda}}, \wp_{\eta_{\grave{A}}} \rangle\} \right), \tag{6.48}$$

where $q \geq 1$.

References

1. B. Batool, M. Ahmad, S. Abdullah, S. Ashraf, R. Chinram, Entropy based Pythagorean probabilistic hesitant fuzzy decision making technique and its application for Fog-Haze factor assessment problem. Entropy **22**, 318 (2020)
2. B. Batool, S.S. Abosuliman, S. Abdullah, S. Ashraf, EDAS method for decision support modeling under the Pythagorean probabilistic hesitant fuzzy aggregation information. J. Ambient Intell. Humanized Comput. **12**, 1–14 (2021)
3. J. Chen, X. Huang, Dual hesitant fuzzy probability. Symmetry **9**, 52 (2017)
4. B. Farhadinia, E. Herrera-viedma, A modification of probabilistic hesitant fuzzy sets and its application to multiple criteria decision making. Iranian J. Fuzzy Syst. **17**, 151–166 (2020)
5. H. Garg, G. Kaur, A robust correlation coefficient for probabilistic dual hesitant fuzzy sets and its applications. Neural Comput. Appl. **32**, 8847–8866 (2020)
6. K. Gong, C. Chen, Multiple-attribute decision making based on equivalence consistency under probabilistic linguistic dual hesitant fuzzy environment. Eng. Appl. Artif. Intell. **85**, 393–401 (2019)
7. K. Gong, C. Chen, Y. Wei, The consistency improvement of probabilistic linguistic hesitant fuzzy preference relations and their application in venture capital group decision making. J. Intell. Fuzzy Syst. **37**, 2925–2936 (2019)
8. Z. Hao, Z. Xu, H. Zhao, Z. Su, Probabilistic dual hesitant fuzzy set and its application in risk evaluation. Knowl.-Based Syst. **127**, 16–28 (2017)
9. F. Jiang, Q. Ma, Multi-attribute group decision making under probabilistic hesitant fuzzy environment with application to evaluate the transformation efficiency. Appl. Intell. **48**, 953–965 (2017). https://doi.org/10.1007/s10489-017-1041-x
10. D.K. Joshi, I. Beg, S. Kumar, Hesitant probabilistic fuzzy linguistic sets with applications in multi-criteria group decision making problems. Mathematics **6**, 47 (2018)
11. L. Li, H. Lei, J. Wang, Q-rung probabilistic dual hesitant fuzzy sets and their application in multi-attribute decision-making. Mathematics **8**, 9 (2020)
12. Q. Pang, H. Wang, Z. Xu, Probabilistic linguistic term sets in multi-attribute group decision making. Inf. Sci. **369**, 128–143 (2016)
13. Z. Ren, Z. Xu, H. Wang, An extended TODIM method under probabilistic dual hesitant fuzzy information and its application on enterprise strategic assessment, in *2017 IEEE International Conference on Industrial Engineering and Engineering Management (IEEM)* (IEEE, Piscataway, 2017), pp. 1464–1468
14. Z. Ren, Z. Xu, H. Wang, The strategy selection problem on artificial intelligence with an integrated VIKOR and AHP method under probabilistic dual hesitant fuzzy information. IEEE Access **7**, 103979–103999 (2019)
15. S.T. Shao, X. Zhang, Measures of probabilistic neutrosophic hesitant fuzzy sets and the application in reducing unnecessary evaluation processes. Mathematics **7**, 649 (2019)

16. S.T. Shao, X.H. Zhang, Y. Li, C.X. Bo, Probabilistic single-valued (interval) neutrosophic hesitant fuzzy set and its application in multi-attribute decision making, Symmetry **10**, 419–439 (2018)
17. S.T. Shao, X.H. Zhang, Q. Zhao, Multi-attribute decision-making based on probabilistic neutrosophic hesitant fuzzy Choquet aggregation operators. Symmetry **11**, 623 (2019)
18. S.T. Shao, X.H. Zhang, Q. Zhao, Multi-attribute decision making based on probabilistic neutrosophic hesitant fuzzy choquet aggregation operators. Symmetry **11**, 623 (2019)
19. C. Song, Z. Xu, H. Zhao, A novel comparison of probabilistic hesitant fuzzy elements in multi-criteria decision making. Symmetry **10**, 177 (2018)
20. S. Xian, H. Guo, J. Chai, W. Wan, Interval probability hesitant fuzzy linguistic analytic hierarchy process and its application in talent selection. J. Intell. Fuzzy Syst. **39**, 2627–2645 (2020)
21. Z. Xu, W. Zhou, Consensus building with a group of decision makers under the hesitant probabilistic fuzzy environment. Fuzzy Optim. Decis. Making **16**, 1–23 (2016)
22. K. Zhao, X. Huang, Probabilistic linguistic hesitant fuzzy set and its application in multiple attribute decision-making. J. Nanchang University **2017**, 1–10 (2017)
23. Y. Zhai, Z. Xu, H. Liao, Measures of probabilistic interval-valued intuitionistic hesitant fuzzy sets and the application in reducing excessive medical examinations. IEEE Trans. Fuzzy Syst. **1**, 1–14 (2017)
24. S. Zhang, Z. Xu, Y. He, Operations and integrations of probabilistic hesitant fuzzy information in decision making. Inf. Fusion **38**, 1–11 (2017)
25. W. Zhou, Z. Xu, Group consistency and group decision making under uncertain probabilistic hesitant fuzzy preference environment. Inf. Sci. **414**, 276–288 (2017)
26. B. Zhu, *Decision Method for Research and Application Based on Preference Relation* (Southeast University, Nanjing, 2014)

Chapter 7
Type 2 Hesitant Fuzzy Set

Abstract In this chapter, we are firstly going to present the notion of type 2 hesitant fuzzy set. Then, the interval-valued type 2 hesitant fuzzy set, which is also known as interval type 2 hesitant fuzzy set, will be given in the next part of this chapter.

7.1 Type 2 Hesitant Fuzzy Set

Feng et al. [3] presented the notion of type 2 hesitant fuzzy set as a family of subsets of the Cartesian product $[0, 1] \times [0, 1]$. Then, they investigated an ordering relation among type 2 hesitant fuzzy sets. In the sequel, Ozlu and Karaaslan [5] mentioned the limitations of Feng et al.'s [3] ordering relation, and they gave a modified type 2 hesitant fuzzy set ordering relation.

Definition 7.1 ([3]) Let X be a reference set. A type 2 hesitant fuzzy set (T2HFS) A on X is defined in terms of a function $h_A(x)$ as follows:

$$A = \{\langle x, h_A(x)\rangle \mid x \in X\}, \tag{7.1}$$

where $h_A(x)$ contains some different values in $[0, 1] \times [0, 1]$ representing a set of determined number of dual membership functions as

$$h_A(x) = \bigcup_{\langle \gamma_A(x), \psi(\gamma_A(x))\rangle \in h_A(x)} \{\langle \gamma_A(x), \psi(\gamma_A(x))\rangle\}. \tag{7.2}$$

The empty and full T2HFSs are defined in the forms of

$$h_A^0(x) = \{\langle 0, 0\rangle\}, \tag{7.3}$$

$$h_A^1(x) = \{\langle 1, 1\rangle\}, \tag{7.4}$$

for any $x \in X$.

B. Farhadinia, *Hesitant Fuzzy Set*, Computational Intelligence Methods and Applications, https://doi.org/10.1007/978-981-16-7301-6_7

To facilitate the comparison rule of T2HFSs, the comparison of two-dimensional coordinate points is taken into consideration as follows (see [3]):

- $(x_1, y_1) \leq (x_2, y_2)$ if and only if $x_1 \leq x_2$ and $y_1 \leq y_2$;
- $\max\{(x_1, y_1), (x_2, y_2)\} = (\max\{x_1, x_2\}, \max\{y_1, y_2\})$.

For the sake of simplicity, $h_A(x)$ is called the type 2 hesitant fuzzy element (T2HFE).

Remark 7.1 Throughout this book, the set of all T2HFSs on the reference set X is denoted by $\mathbb{T2HFS}(X)$.

Given a T2HFS, the concept of lower, upper, α-lower, and α-upper bounds are represented as:

- Lower bound: $h_A^-(x) = \bigcup_{\langle \gamma_{A^-}(x), \psi(\gamma_{A^-}(x))\rangle \in h_A^-(x)} \{\langle \min\{\gamma_A(x)\}, \min\{\psi(\gamma_A(x))\}\rangle\}$;
- Upper bound: $h_A^+(x) = \bigcup_{\langle \gamma_{A^+}(x), \psi(\gamma_{A^+}(x))\rangle \in h_A^+(x)} \{\langle \max\{\gamma_A(x)\}, \max\{\psi(\gamma_A(x))\}\rangle\}$;
- $[\alpha, \beta]$-Lower bound: $h_{A[\alpha,\beta]}^-(x) = \bigcup_{\langle \gamma_A(x), \psi(\gamma_A(x))\rangle \in h_A(x)} \{\langle \gamma_A(x), \psi(\gamma_A(x))\rangle \mid \gamma_A(x) \leq \alpha, \psi(\gamma_A(x)) \leq \beta\}$;
- $[\alpha, \beta]$-Upper bound: $h_{A[\alpha,\beta]}^+(x) = \bigcup_{\langle \gamma_A(x), \psi(\gamma_A(x))\rangle \in h_A(x)} \{\langle \gamma_A(x), \psi(\gamma_A(x))\rangle \mid \gamma_A(x) \geq \alpha, \psi(\gamma_A(x)) \geq \beta\}$.

Example 7.1 Let $X = \{x\}$ be the reference set, $h_A(x) = \bigcup_{\langle \gamma_A(x), \psi(\gamma_A(x))\rangle \in h_A(x)} \{\langle \gamma_A(x), \psi(\gamma_A(x))\rangle\}$ $= \bigcup_{\langle \gamma_A(x), \psi(\gamma_A(x))\rangle \in h_A(x)} \{\langle 0.2, 0.5\rangle, \langle 0.4, 0.6\rangle, \langle 0.6, 0.4\rangle\}$ be the T2HFEs of x in the set A. Then, we have $h_A^-(x) = \{\langle 0.2, 0.4\}\rangle\}$, $h_A^+(x) = \{\langle 0.6, 0.6\}\rangle\}$, $h_{A[0.3,0.5]}^-(x) = \{\langle 0.2, 0.5\}\rangle\}$, and $h_{A[0.3,0.5]}^+(x) = \{\langle 0.4, 0.6\}\rangle\}$.

For the T2HFEs $h_A = \bigcup_{\langle \gamma_A, \psi(\gamma_A)\rangle \in h_A} \{\langle \gamma_A, \psi(\gamma_A)\rangle\}$, $h_{A_1} = \bigcup_{\langle \gamma_{A_1}, \psi(\gamma_{A_1})\rangle \in h_{A_1}} \{\langle \gamma_{A_1}, \psi(\gamma_{A_1})\rangle\}$ and $h_{A_2} = \bigcup_{\langle \gamma_{A_2}, \psi(\gamma_{A_2})\rangle \in h_{A_2}} \{\langle \gamma_{A_2}, \psi(\gamma_{A_2})\rangle\}$ the following operations are defined (see [3]):

$$h_{A^c} = \bigcup_{\langle \gamma_A, \psi(\gamma_A)\rangle \in h_A} \{\langle 1 - \gamma_A, 1 - \psi(\gamma_A)\rangle\}; \tag{7.5}$$

$$h_{A_1} \cup h_{A_2} = \bigcup_{\langle \gamma^{\cup}, \psi(\gamma^{\cup})\rangle \in h_{A_1} \cup h_{A_2}} \{\langle \gamma^{\cup}, \psi(\gamma^{\cup})\rangle \mid \langle \gamma^{\cup}, \psi(\gamma^{\cup})\rangle \geq \max\{h_{A_1}^-, h_{A_2}^-\}\}; \tag{7.6}$$

$$h_{A_1} \cap h_{A_2} = \bigcup_{\langle \gamma^{\cap}, \psi(\gamma^{\cap})\rangle \in h_{A_1} \cap h_{A_2}} \{\langle \gamma^{\cap}, \psi(\gamma^{\cap})\rangle \mid \langle \gamma^{\cap}, \psi(\gamma^{\cap})\rangle \leq \min\{h_{A_1}^+, h_{A_2}^+\}\}. \tag{7.7}$$

Example 7.2 Let $X = \{x\}$ be the reference set, $h_{A_1}(x) = \bigcup_{\langle \gamma_{A_1}(x), \psi(\gamma_{A_1}(x))\rangle \in h_{A_1}} \{\langle \gamma_{A_1}(x), \psi(\gamma_{A_1}(x))\rangle\} = \{\langle 0.2, 0.5\rangle, \langle 0.4, 0.6\rangle, \langle 0.6, 0.4\rangle\}$ and $h_{A_2}(x) = \bigcup_{\langle \gamma_{A_2}(x), \psi(\gamma_{A_2}(x))\rangle \in h_{A_2}} \{\langle \gamma_{A_2}(x), \psi(\gamma_{A_2}(x))\rangle\} = \{\langle 0.3, 0.4\rangle, \langle 0.4, 0.5\rangle\}$ be two T2HFEs. Then, we achieve that $h^-_{A_1}(x) = \{\langle 0.2, 0.4\}\rangle\}$, $h^+_{A_1}(x) = \{\langle 0.6, 0.6\}\rangle\}$, $h^-_{A_2}(x) = \{\langle 0.3, 0.4\}\rangle\}$, and $h^+_{A_2}(x) = \{\langle 0.4, 0.5\}\rangle\}$. Therefore,

$$\max\{h^-_{A_1}, h^-_{A_2}\} = \{\langle 0.3, 0.4\}\rangle\};$$
$$\min\{h^+_{A_1}, h^+_{A_2}\} = \{\langle 0.4, 0.5\}\rangle\}.$$

Keeping the set operations in the mind, we get

$$h_{A_1^c}(x) = \{\langle 0.8, 0.5\rangle, \langle 0.6, 0.4\rangle, \langle 0.4, 0.6\rangle\};$$
$$h_{A_1} \cup h_{A_2} = \{\langle 0.4, 0.6\rangle, \langle 0.6, 0.4\rangle, \langle 0.3, 0.4\rangle, \langle 0.4, 0.5\rangle\};$$
$$h_{A_1} \cap h_{A_2} = \{\langle 0.2, 0.5\rangle, \langle 0.3, 0.4\rangle, \langle 0.4, 0.5\rangle\}.$$

7.2 Interval Type 2 Hesitant Fuzzy Set

Hu et al. [4] presented the concept of interval type-2 hesitant fuzzy set as a generalization of interval type-2 fuzzy set and hesitant fuzzy set. Then, they developed a number of interval type-2 hesitant fuzzy set operation laws, score function, and aggregation operators. Deveci [2] investigated a multiple criteria decision making technique under interval type-2 hesitant fuzzy environment for assessing the quality service of airline companies.

By taking the reference set X into account, a trapezoidal interval Type 2 fuzzy number (TIT2FN) $\widetilde{A}$ on X is defined in terms of two lower and upper trapezoidal membership functions as (see [1]):

$$\widetilde{A} = \left\{\left\langle x, u^L_{\widetilde{A}}(x), u^U_{\widetilde{A}}(x)\right\rangle \mid x \in X\right\}, \tag{7.8}$$

where $u^L_{\widetilde{A}}(x) = (u^L_{1\widetilde{A}}(x), u^L_{2\widetilde{A}}(x), u^L_{3\widetilde{A}}(x), u^L_{4\widetilde{A}}(x); \omega^l_{u^L_{\widetilde{A}}(x)}, \omega^u_{u^L_{\widetilde{A}}(x)})$ and $u^U_{\widetilde{A}}(x) = (u^U_{1\widetilde{A}}(x), u^U_{2\widetilde{A}}(x), u^U_{3\widetilde{A}}(x), u^U_{4\widetilde{A}}(x); \omega^l_{u^U_{\widetilde{A}}(x)}, \omega^u_{u^U_{\widetilde{A}}(x)})$ are type 1 fuzzy sets such that $\omega^l_{u^L_{\widetilde{A}}(x)}, \omega^u_{u^L_{\widetilde{A}}(x)}, \omega^l_{u^U_{\widetilde{A}}(x)}$ and $\omega^u_{u^U_{\widetilde{A}}(x)}$ are all belong to the interval [0, 1].

Definition 7.2 ([4]) Let X be a reference set. An interval type 2 hesitant fuzzy set (IT2HFS) $\widetilde{A}$ on X is defined in terms of a functions $h_{\widetilde{A}}(x)$ as follows:

$$\widetilde{A} = \{\langle x, h_{\widetilde{A}}(x)\rangle \mid x \in X\}, \tag{7.9}$$

where $h_{\widetilde{A}}(x)$ contains some different TIT2FNs as

$$\begin{aligned} h_{\widetilde{A}}(x) &= \bigcup_{\left\langle u^L_{\widetilde{A}}(x), u^U_{\widetilde{A}}(x)\right\rangle \in h_{\widetilde{A}}(x)} \left\{\left\langle u^L_{\widetilde{A}}(x), u^U_{\widetilde{A}}(x)\right\rangle\right\} \\ &= \bigcup_{\left\langle u^L_{\widetilde{A}}(x), u^U_{\widetilde{A}}(x)\right\rangle \in h_{\widetilde{A}}(x)} \left\{\left\langle\left(u^L_{1\widetilde{A}}(x), u^L_{2\widetilde{A}}(x), u^L_{3\widetilde{A}}(x), u^L_{4\widetilde{A}}(x); \omega^l_{u^L_{\widetilde{A}}(x)}, \omega^u_{u^L_{\widetilde{A}}(x)}\right),\right.\right. \\ &\left.\left.\left(u^U_{1\widetilde{A}}(x), u^U_{2\widetilde{A}}(x), u^U_{3\widetilde{A}}(x), u^U_{4\widetilde{A}}(x); \omega^l_{u^U_{\widetilde{A}}(x)}, \omega^u_{u^U_{\widetilde{A}}(x)}\right)\right\rangle\right\}. \end{aligned} \tag{7.10}$$

For the sake of simplicity, $h_{\widetilde{A}}(x)$ is called the interval type 2 hesitant fuzzy element (IT2HFE).

Remark 7.2 Throughout this book, the set of all IT2HFSs on the reference set X is denoted by $\mathbb{IT2HFS}(X)$.

For the IT2HFEs $h_{\widetilde{A}} = \bigcup_{\langle u^L_{\widetilde{A}}, u^U_{\widetilde{A}}\rangle \in h_{\widetilde{A}}} \{\langle (u^L_{1\widetilde{A}}, u^L_{2\widetilde{A}}, u^L_{3\widetilde{A}}, u^L_{4\widetilde{A}}; \omega^l_{u^L_{\widetilde{A}}}, \omega^u_{u^L_{\widetilde{A}}}),$ $(u^U_{1\widetilde{A}}, u^U_{2\widetilde{A}}, u^U_{3\widetilde{A}}, u^U_{4\widetilde{A}}; \omega^l_{u^U_{\widetilde{A}}}, \omega^u_{u^U_{\widetilde{A}}})\rangle\}$, $h_{\widetilde{A}_1} = \bigcup_{\langle u^L_{\widetilde{A}_1}, u^U_{\widetilde{A}_1}\rangle \in h_{\widetilde{A}_1}} \{\langle (u^L_{1\widetilde{A}_1}, u^L_{2\widetilde{A}_1}, u^L_{3\widetilde{A}_1}, u^L_{4\widetilde{A}_1}; \omega^l_{u^L_{\widetilde{A}_1}}, \omega^u_{u^L_{\widetilde{A}_1}}),$ $(u^U_{1\widetilde{A}_1}, u^U_{2\widetilde{A}_1}, u^U_{3\widetilde{A}_1}, u^U_{4\widetilde{A}_1}; \omega^l_{u^U_{\widetilde{A}_1}}, \omega^u_{u^U_{\widetilde{A}_1}})\rangle\}$ and $h_{\widetilde{A}_2} = \bigcup_{\langle u^L_{\widetilde{A}_2}, u^U_{\widetilde{A}_2}\rangle \in h_{\widetilde{A}_2}} \{\langle (u^L_{1\widetilde{A}_2}, u^L_{2\widetilde{A}_2}, u^L_{3\widetilde{A}_2}, u^L_{4\widetilde{A}_2}; \omega^l_{u^L_{\widetilde{A}_2}}, \omega^u_{u^L_{\widetilde{A}_2}}),$ $(u^U_{1\widetilde{A}_2}, u^U_{2\widetilde{A}_2}, u^U_{3\widetilde{A}_2}, u^U_{4\widetilde{A}_2}; \omega^l_{u^U_{\widetilde{A}_2}}, \omega^u_{u^U_{\widetilde{A}_2}})\rangle\}$ the following operations are defined (see [4]):

$$\begin{aligned} h_{\widetilde{A}_1} \oplus h_{\widetilde{A}_2} = \bigcup_{\substack{\left\langle u^L_{\widetilde{A}_1}, u^U_{\widetilde{A}_1}\right\rangle \in h_{\widetilde{A}_1} \\ \left\langle u^L_{\widetilde{A}_2}, u^U_{\widetilde{A}_2}\right\rangle \in h_{\widetilde{A}_2}}} & \left\{\left\langle\left(u^L_{1\widetilde{A}_1} + u^L_{1\widetilde{A}_2} - u^L_{1\widetilde{A}_1} \times u^L_{1\widetilde{A}_2},\right.\right.\right. \\ & u^L_{2\widetilde{A}_1} + u^L_{2\widetilde{A}_2} - u^L_{2\widetilde{A}_1} \times u^L_{2\widetilde{A}_2}, \\ & u^L_{3\widetilde{A}_1} + u^L_{3\widetilde{A}_2} - u^L_{3\widetilde{A}_1} \times u^L_{3\widetilde{A}_2}, \\ & u^L_{4\widetilde{A}_1} + u^L_{4\widetilde{A}_2} - u^L_{4\widetilde{A}_1} \times u^L_{4\widetilde{A}_2}; \min\left\{\omega^l_{u^L_{\widetilde{A}_1}}, \omega^l_{u^L_{\widetilde{A}_2}}\right\}, \end{aligned}$$

$$\min\left\{\omega^u_{u^L_{\widetilde{A}_1}}, \omega^u_{u^L_{\widetilde{A}_2}}\right\}\Big),$$

$$\Big(u^U_{1\widetilde{A}_1} + u^U_{1\widetilde{A}_2} - u^U_{1\widetilde{A}_1} \times u^U_{1\widetilde{A}_2},$$

$$u^U_{2\widetilde{A}_1} + u^U_{2\widetilde{A}_2} - u^U_{2\widetilde{A}_1} \times u^U_{2\widetilde{A}_2},$$

$$u^U_{3\widetilde{A}_1} + u^U_{3\widetilde{A}_2} - u^U_{3\widetilde{A}_1} \times u^U_{3\widetilde{A}_2},$$

$$u^U_{4\widetilde{A}_1} + u^U_{4\widetilde{A}_2} - u^U_{4\widetilde{A}_1} \times u^U_{4\widetilde{A}_2}; \min\left\{\omega^l_{u^U_{\widetilde{A}_1}}, \omega^l_{u^U_{\widetilde{A}_2}}\right\},$$

$$\min\left\{\omega^u_{u^U_{\widetilde{A}_1}}, \omega^u_{u^U_{\widetilde{A}_2}}\right\}\Big)\Big\rangle\Big\}; \tag{7.11}$$

$$h_{\widetilde{A}_1} \otimes h_{\widetilde{A}_2} = \bigcup_{\substack{\left\langle u^L_{\widetilde{A}_1}, u^U_{\widetilde{A}_1}\right\rangle \in h_{\widetilde{A}_1} \\ \left\langle u^L_{\widetilde{A}_2}, u^U_{\widetilde{A}_2}\right\rangle \in h_{\widetilde{A}_2}}} \Big\{\Big\langle\Big(u^L_{1\widetilde{A}_1} \times u^L_{1\widetilde{A}_2},$$

$$u^L_{2\widetilde{A}_1} \times u^L_{2\widetilde{A}_2},$$

$$u^L_{3\widetilde{A}_1} \times u^L_{3\widetilde{A}_2},$$

$$u^L_{4\widetilde{A}_1} \times u^L_{4\widetilde{A}_2}; \min\left\{\omega^l_{u^L_{\widetilde{A}_1}}, \omega^l_{u^L_{\widetilde{A}_2}}\right\},$$

$$\min\left\{\omega^u_{u^L_{\widetilde{A}_1}}, \omega^u_{u^L_{\widetilde{A}_2}}\right\}\Big),$$

$$\Big(u^U_{1\widetilde{A}_1} \times u^U_{1\widetilde{A}_2},$$

$$u^U_{2\widetilde{A}_1} \times u^U_{2\widetilde{A}_2},$$

$$u^U_{3\widetilde{A}_1} \times u^U_{3\widetilde{A}_2},$$

$$u^U_{4\widetilde{A}_1} \times u^U_{4\widetilde{A}_2}; \min\{\omega^l_{u^U_{\widetilde{A}_1}}, \omega^l_{u^U_{\widetilde{A}_2}}\right\},$$

$$\min\left\{\omega^u_{u^U_{\widetilde{A}_1}}, \omega^u_{u^U_{\widetilde{A}_2}}\right\}\Big)\Big\rangle\Big\}; \tag{7.12}$$

$$\begin{aligned}
\lambda h_{\tilde{A}} = \bigcup_{\left\langle u^L_{\tilde{A}}, u^U_{\tilde{A}} \right\rangle \in h_{\tilde{A}}} & \Big\{\Big\langle \Big(1-(1-u^L_{1\tilde{A}})^{\lambda}, \\
& 1-\left(1-u^L_{2\tilde{A}}\right)^{\lambda}, \\
& 1-\left(1-u^L_{3\tilde{A}}\right)^{\lambda}, \\
& 1-\left(1-u^L_{4\tilde{A}}\right)^{\lambda}; \omega^l_{u^L_{\tilde{A}}}, \omega^u_{u^L_{\tilde{A}}}\Big), \\
& \Big(1-(1-u^U_{1\tilde{A}})^{\lambda}, \\
& 1-\left(1-u^U_{2\tilde{A}}\right)^{\lambda}, \\
& 1-\left(1-u^U_{3\tilde{A}}\right)^{\lambda}, \\
& 1-\left(1-u^U_{4\tilde{A}}\right)^{\lambda}; \omega^l_{u^U_{\tilde{A}}}, \omega^u_{u^U_{\tilde{A}}}\Big)\Big\rangle\Big\};
\end{aligned} \tag{7.13}$$

$$\begin{aligned}
h_{\tilde{A}}^{\lambda} = \bigcup_{\langle u^L_{\tilde{A}}, u^U_{\tilde{A}} \rangle \in h_{\tilde{A}}} & \Big\{\Big\langle \Big((u^L_{1\tilde{A}})^{\lambda}, \\
& \left(u^L_{2\tilde{A}}\right)^{\lambda}, \\
& \left(u^L_{3\tilde{A}}\right)^{\lambda}, \\
& \left(u^L_{4\tilde{A}}\right)^{\lambda}; \omega^l_{u^L_{\tilde{A}}}, \omega^u_{u^L_{\tilde{A}}}\Big), \\
& \Big((u^U_{1\tilde{A}})^{\lambda}, \\
& \left(u^U_{2\tilde{A}}\right)^{\lambda}, \\
& \left(u^U_{3\tilde{A}}\right)^{\lambda}, \\
& \left(u^U_{4\tilde{A}}\right)^{\lambda}; \omega^l_{u^U_{\tilde{A}}}, \omega^u_{u^U_{\tilde{A}}}\Big)\Big\rangle\Big\},
\end{aligned} \tag{7.14}$$

for any $\lambda > 0$.

The other kinds of the above definitions based on various T-norms and S-norms can be found in [4].

References

1. S.M. Chen, L.W. Lee, Fuzzy multiple attributes group decision-making based on the interval type-2 TOPSIS method. Expert Syst. Appl. **37**, 2790–2798 (2010)
2. M. Deveci, E. Ozcan, R. John, S. Ceren Oner, Interval type-2 hesitant fuzzy set method for improving the service quality of domestic airlines in Turkey. J. Air Trans. Manag. **69**, 83–98 (2018)
3. L. Feng, F. Chuan-qiang, X. Wei-he, Type-2 hesitant fuzzy sets. Fuzzy Inf. Eng. **10**, 249–259 (2018)
4. J. Hu, K. Xiao, X. Chen, Y. Liu, Interval type-2 hesitant fuzzy set and its application in multi-criteria decision making. Comput. Ind. Eng. **87**, 91–103 (2015)
5. S. Ozlu, F. Karaaslan, Some distance measures for Type-2 hesitant fuzzy sets and their application to multi-criteria group decision making problems. Soft Comput. **24**, 9965–9980 (2020)

Chapter 8
Hesitant Bipolar Fuzzy Set

Abstract In this chapter, we first introduce the concept of hesitant bipolar fuzzy set. Then, we deal with the concept that comes from bipolar concept, and it is stated more or less different from the concept of hesitant bipolar fuzzy set. Indeed, we are going to introduce the concept of hesitant bipolar-valued fuzzy set. In the subsequent part of this chapter, we will represent the concept of hesitant bipolar-valued neutrosophic set.

8.1 Hesitant Bipolar Fuzzy Set

Wei et al. [5] introduced a set of aggregation operators for aggregating hesitant bipolar fuzzy sets and utilized them in hesitant bipolar fuzzy multiple criteria decision making problems. Since the concept of hesitant fuzzy set cannot accommodate incompatible bipolarity, Han et al. [4] represented the concept of hesitant bipolar fuzzy set involved in a multiple criteria decision making technique which has been enhanced by the adoption of TOPSIS approach. Xu and Wei [6] presented the concept of dual hesitant bipolar fuzzy set, which is indeed nothing else, except the concept of hesitant bipolar fuzzy set, and developed a class of hesitant bipolar fuzzy set aggregation operators. Gao et al. [2] in the same line of Xu and Wei [6] investigated multiple criteria decision making problems by using the Hamacher aggregation operators under the dual hesitant bipolar fuzzy environment.

Before dealing with the main concepts of combination of bipolar and hesitant fuzzy notions, let us introduce the concept of bipolar fuzzy set: keeping the reference set X in mind, a bipolar fuzzy set (BFS) A on X is defined in terms of two functions u_A and v_A as follows [7]:

$$A = \{\langle x, (u_A(x), v_A(x))\rangle \mid x \in X\}, \tag{8.1}$$

B. Farhadinia, *Hesitant Fuzzy Set*, Computational Intelligence Methods
and Applications, https://doi.org/10.1007/978-981-16-7301-6_8

where $u_A(x)$ and $v_A(x)$ are, respectively, the sets of some different values in [0, 1] and $[-1, 0]$, and moreover, they represent positive membership degree and negative membership degree of the element $x \in X$ to A, respectively. The positive and negative membership degrees satisfy

$$0 \leq u_A(x) \leq 1, \quad -1 \leq v_A(x) \leq 0. \tag{8.2}$$

Given a fixed $x \in X$, some operational laws on BFSs are defined as follows [3]:

$$\begin{aligned}
A^c &= \{\langle x, (1 - u_A(x), |v_A(x)| - 1)\rangle \mid x \in X\};\\
A_1 \cup A_2 &= \{\langle x, (\max\{u_{A_1}(x), u_{A_2}(x)\}, \min\{v_{A_1}(x), v_{A_2}(x)\})\rangle \mid x \in X\};\\
A_1 \cap A_2 &= \{\langle x, (\min\{u_{A_1}(x), u_{A_2}(x)\}, \max\{v_{A_1}(x), v_{A_2}(x)\})\rangle \mid x \in X\};\\
A_1 \oplus A_2 &= \{\langle x, (u_{A_1}(x) + u_{A_2}(x) - u_{A_1}(x) \times u_{A_2}(x),\\
&\quad - |v_{A_1}(x)| \times |v_{A_2}(x)|)\rangle \mid x \in X\};\\
A_1 \otimes A_2 &= \{\langle x, (u_{A_1}(x) \times u_{A_2}(x), v_{A_1}(x) + v_{A_2}(x)\\
&\quad - v_{A_1}(x) \times v_{A_2}(x))\rangle \mid x \in X\};\\
\lambda A &= \{\langle x, (1 - (1 - u_A(x))^\lambda, -|v_A(x)|^\lambda)\rangle \mid x \in X\}, \quad \lambda > 0;\\
A^\lambda &= \{\langle x, (|u_A(x)|^\lambda, -1 + |1 + v_A(x)|^\lambda)\rangle \mid x \in X\}, \quad \lambda > 0.
\end{aligned}$$

Example 8.1 Let $X = \{x_1, x_2\}$ be the reference set, $(u_A(x_1), v_A(x_1)) = \{(0.2, -0.3)\}$ and $(u_A(x_2), v_A(x_2)) = \{(0.4, -0.8)\}$. With respect to these two terms, the set A can be considered as an BFS where

$$A = \{\langle x_1, (0.2, -0.3)\rangle, \langle x_2, (0.4, -0.8)\rangle\}.$$

Definition 8.1 ([5]) Let X be a reference set. A bipolar hesitant fuzzy set (BHFS) A on X is defined in terms of two functions $u_A(x)$ and $v_A(x)$ as follows:

$$A = \{\langle x, u_A(x), v_A(x)\rangle \mid x \in X\},$$

where the positive membership degree $u_A(x) \in [0, 1]$ indicates the possible satisfaction degree of an element $x \in X$ to the property corresponding to A, and the negative membership degree $v_A(x) \in [-1, 0]$ indicates the possible satisfaction degree of an element $x \in X$ to some implicit counter property corresponding to A.

Here, for all $x \in X$, if we consider $u_A(x) = \bigcup_{\gamma_A \in u_A(x)}\{\gamma_A\}$ and $v_A(x) = \bigcup_{\eta_A \in v_A(x)}\{\eta_A\}$, then we have

$$0 \leq \gamma_A \leq 1, \quad -1 \leq \eta_A \leq 0.$$

For the sake of simplicity, the pair $h_A(x) = (u_A(x), v_A(x))$ is called the bipolar hesitant fuzzy element (BHFE).

Notice that $h_A = (u_A, v_A)$ stands here for $h_A = \bigcup_{\substack{\gamma_A \in u_A \\ \eta_A \in v_A}} \{(\gamma_A, \eta_A)\}$.

Remark 8.1 Throughout this book, the set of all BHFSs on the reference set X is denoted by $\mathbb{BHFS}(X)$.

Example 8.2 Let $X = \{x_1, x_2\}$ be the reference set, $h_A(x_1) = (u_A(x_1), v_A(x_1)) = \{(0.2, -0.5), (0.2, -0.3)\}$ and $h_A(x_2) = (u_A(x_2), v_A(x_2)) = \{(0.3, -0.4), (0.1, -0.6), (0.1, -0.8)\})$, then $h_A(x_i)$ for $i = 1, 2$ are the BHFEs of x_i $(i = 1, 2)$ in the set A. Thus, A is a BHFS, and it is denoted by

$$
\begin{aligned}
A = \{ & \langle x_1, \{(0.2, -0.5), (0.2, -0.3)\}, \rangle, \\
& \langle x_2, \{(0.3, -0.4), (0.1, -0.6), (0.1, -0.8)\}\rangle\}.
\end{aligned}
$$

For the BHFEs $h_A = (u_A, v_A)$, $h_{A_1} = (u_{A_1}, v_{A_1})$, and $h_{A_2} = (u_{A_2}, v_{A_2})$ the following operations are defined (see [5]):

$$
\begin{aligned}
h_{A_1} \oplus h_{A_2} &= (u_{A_1} \oplus_B u_{A_2}, v_{A_1} \otimes_B v_{A_2}) \\
&= \bigcup_{\substack{\gamma_{A_1} \in u_{A_1}, \eta_{A_1} \in v_{A_1} \\ \gamma_{A_2} \in u_{A_2}, \eta_{A_2} \in v_{A_2}}} \{(\gamma_{A_1} + \gamma_{A_2} - \gamma_{A_1} \times \gamma_{A_2}, -|\eta_{A_1}||\eta_{A_2}|)\}; \qquad (8.3)
\end{aligned}
$$

$$
\begin{aligned}
h_{A_1} \otimes h_{A_2} &= (u_{A_1} \otimes_B u_{A_2}, v_{A_1} \oplus_B v_{A_2}) \\
&= \bigcup_{\substack{\gamma_{A_1} \in u_{A_1}, \eta_{A_1} \in v_{A_1} \\ \gamma_{A_2} \in u_{A_2}, \eta_{A_2} \in v_{A_2}}} \{(\gamma_{A_1} \times \gamma_{A_2}, \eta_{A_1} + \eta_{A_2} - \eta_{A_1} \times \eta_{A_2})\}, \qquad (8.4)
\end{aligned}
$$

$$
\lambda h_A = \bigcup_{\gamma_A \in u_A, \eta_A \in v_A} \{(1 - (1 - \gamma_A)^\lambda, -|\eta_A|^\lambda)\}, \qquad (8.5)
$$

$$
(h_A)^\lambda = \bigcup_{\gamma_A \in u_A, \eta_A \in v_A} \{({\gamma_A}^\lambda, -1 + |1 + \eta_A|^\lambda)\}, \qquad (8.6)
$$

where

$$
\begin{aligned}
0 \le \gamma_A \le 1, &\quad -1 \le \eta_A \le 0, \\
0 \le \gamma_{A_1} \le 1, &\quad -1 \le \eta_{A_1} \le 0, \\
0 \le \gamma_{A_2} \le 1, &\quad -1 \le \eta_{A_2} \le 0.
\end{aligned}
$$

8.2 Hesitant Bipolar-Valued Fuzzy Set

Mandal and Ranadive [2] presented two distinguished aspects: one that defines the concept of hesitant bipolar-valued fuzzy set, and the other that defines the concept of bipolar-valued hesitant fuzzy set. The former concept is the combination of the concept of bipolar-valued fuzzy set and hesitant fuzzy set. The latter concept extends the concept of hesitant fuzzy set, by injecting the concept of bipolar-valued fuzzy set. The fundamental characteristic of hesitant bipolar-valued fuzzy set is the values of the positive membership function and the negative membership function which are set of positive numbers and negative numbers rather than exact positive number and negative number. The fundamental characteristic of bipolar-valued hesitant fuzzy set is the values of the membership function which is a set of bipolar-valued fuzzy numbers rather than set of exact numbers.

Definition 8.2 ([2]) Let X be a reference set. A hesitant bipolar-valued fuzzy set (HBVFS) $\widetilde{A}$ on X is defined in terms of two functions $u_{\widetilde{A}}(x)$ and $v_{\widetilde{A}}(x)$ as follows:

$$\widetilde{A} = \{\langle x, u_{\widetilde{A}}(x), v_{\widetilde{A}}(x)\rangle \mid x \in X\}, \tag{8.7}$$

where the hesitant fuzzy positive element $u_{\widetilde{A}}(x)$ is a set of some values in $[0, 1]$ indicating the possible satisfaction degree of an element $x \in X$ to the property corresponding to $\widetilde{A}$, and the hesitant fuzzy negative element $v_{\widetilde{A}}(x)$ is a set of some values in $[-1, 0]$ indicating the possible satisfaction degree of an element $x \in X$ to some implicit counter property corresponding to $\widetilde{A}$.

Here, for all $x \in X$, if we take $u_{\widetilde{A}}(x) = \bigcup_{\gamma_{\widetilde{A}} \in u_{\widetilde{A}}(x)} \{\gamma_{\widetilde{A}}\}$ and $v_{\widetilde{A}}(x) = \bigcup_{\eta_{\widetilde{A}} \in v_{\widetilde{A}}(x)} \{\eta_{\widetilde{A}}\}$, then we have

$$0 \leq \gamma_{\widetilde{A}} \leq 1, \quad -1 \leq \eta_{\widetilde{A}} \leq 0.$$

For the sake of simplicity, the pair $h_{\widetilde{A}}(x) = (u_{\widetilde{A}}(x), v_{\widetilde{A}}(x))$ is called the hesitant bipolar-valued fuzzy element (HBVFE).

Remark 8.2 Throughout this book, the set of all HBVFSs on the reference set X is denoted by $\mathbb{HBVFS}(X)$.

Example 8.3 Let $X = \{x_1, x_2\}$ be the reference set, $h_{\widetilde{A}}(x_1) = (u_{\widetilde{A}}(x_1), v_{\widetilde{A}}(x_1)) = (\{0.2\}, \{-0.5 - 0.3\})$ and $h_{\widetilde{A}}(x_2) = (u_{\widetilde{A}}(x_2), v_{\widetilde{A}}(x_2)) = (\{0.3, 0.1\}, \{-0.4, -0.6, -0.8\})$, then $h_{\widetilde{A}}(x_i)$ for $i = 1, 2$ are the HBVFEs of x_i $(i = 1, 2)$ in the set $\widetilde{A}$. Thus, $\widetilde{A}$ is a HBVFS, and it is denoted by

$$\widetilde{A} = \{\langle x_1, (\{0.2\}, \{-0.5, -0.3\})\rangle, \langle x_2, (\{0.3, 0.1\}, \{-0.4, -0.6, -0.8\})\rangle\}.$$

For the HBVFEs $h_{\widetilde{A}} = (u_{\widetilde{A}}, v_{\widetilde{A}})$, $h_{\widetilde{A}_1} = (u_{\widetilde{A}_1}, v_{\widetilde{A}_1})$, and $h_{\widetilde{A}_2} = (u_{\widetilde{A}_2}, v_{\widetilde{A}_2})$ the following operations are defined (see [5]):

$$(h_{\widetilde{A}})^c = \Big(\bigcup_{\gamma_{\widetilde{A}} \in u_{\widetilde{A}}} \{1 - \gamma_{\widetilde{A}}\}, \bigcup_{\eta_{\widetilde{A}} \in v_{\widetilde{A}}} \{-1 - \eta_{\widetilde{A}}\} \Big); \tag{8.8}$$

$$\begin{aligned} h_{\widetilde{A}_1} \cup h_{\widetilde{A}_2} &= (u_{\widetilde{A}_1} \cup u_{\widetilde{A}_2}, v_{\widetilde{A}_1} \cap v_{\widetilde{A}_2}) \\ &= \Big(\bigcup_{\gamma_{\widetilde{A}_1} \in u_{\widetilde{A}_1}, \gamma_{\widetilde{A}_2} \in u_{\widetilde{A}_2}} \{\max\{\gamma_{\widetilde{A}_1}, \gamma_{\widetilde{A}_2}\}\}, \bigcup_{\eta_{\widetilde{A}_1} \in v_{\widetilde{A}_1}, \eta_{\widetilde{A}_2} \in v_{\widetilde{A}_2}} \{\min\{\eta_{\widetilde{A}_1}, \eta_{\widetilde{A}_2}\}\} \Big); \end{aligned} \tag{8.9}$$

$$\begin{aligned} h_{\widetilde{A}_1} \cap h_{\widetilde{A}_2} &= (u_{\widetilde{A}_1} \cap u_{\widetilde{A}_2}, v_{\widetilde{A}_1} \cup v_{\widetilde{A}_2}) \\ &= \Big(\bigcup_{\gamma_{\widetilde{A}_1} \in u_{\widetilde{A}_1}, \gamma_{\widetilde{A}_2} \in u_{\widetilde{A}_2}} \{\min\{\gamma_{\widetilde{A}_1}, \gamma_{\widetilde{A}_2}\}\}, \bigcup_{\eta_{\widetilde{A}_1} \in v_{\widetilde{A}_1}, \eta_{\widetilde{A}_2} \in v_{\widetilde{A}_2}} \{\max\{\eta_{\widetilde{A}_1}, \eta_{\widetilde{A}_2}\}\} \Big); \end{aligned} \tag{8.10}$$

$$\begin{aligned} h_{\widetilde{A}_1} \oplus h_{\widetilde{A}_2} &= (u_{\widetilde{A}_1} \oplus u_{\widetilde{A}_2}, v_{\widetilde{A}_1} \oplus v_{\widetilde{A}_2}) \\ &= \Big(\bigcup_{\gamma_{\widetilde{A}_1} \in u_{\widetilde{A}_1}, \gamma_{\widetilde{A}_2} \in u_{\widetilde{A}_2}} \{\gamma_{\widetilde{A}_1} + \gamma_{\widetilde{A}_2} - \gamma_{\widetilde{A}_1} \times \gamma_{\widetilde{A}_2}\}, \\ &\qquad \bigcup_{\eta_{\widetilde{A}_1} \in v_{\widetilde{A}_1}, \eta_{\widetilde{A}_2} \in v_{\widetilde{A}_2}} \{\eta_{\widetilde{A}_1} + \eta_{\widetilde{A}_2} + \eta_{\widetilde{A}_1} \times \eta_{\widetilde{A}_2}\} \Big); \end{aligned} \tag{8.11}$$

$$\begin{aligned} h_{\widetilde{A}_1} \otimes h_{\widetilde{A}_2} &= (u_{\widetilde{A}_1} \otimes u_{\widetilde{A}_2}, v_{\widetilde{A}_1} \otimes v_{\widetilde{A}_2}) \\ &= \Big(\bigcup_{\gamma_{\widetilde{A}_1} \in u_{\widetilde{A}_1}, \gamma_{\widetilde{A}_2} \in u_{\widetilde{A}_2}} \{\gamma_{\widetilde{A}_1} \times \gamma_{\widetilde{A}_2}\}, \bigcup_{\eta_{\widetilde{A}_1} \in v_{\widetilde{A}_1}, \eta_{\widetilde{A}_2} \in v_{\widetilde{A}_2}} \{-\eta_{\widetilde{A}_1} \times \eta_{\widetilde{A}_2}\} \Big); \end{aligned} \tag{8.12}$$

$$\lambda h_{\widetilde{A}} = \Big(\bigcup_{\gamma_{\widetilde{A}} \in u_{\widetilde{A}}} \{1 - (1 - \gamma_{\widetilde{A}})^{\lambda}\}, \bigcup_{\eta_{\widetilde{A}} \in v_{\widetilde{A}}} \{-1 + (1 + \eta_{\widetilde{A}})^{\lambda}\} \Big); \tag{8.13}$$

$$(h_{\widetilde{A}})^{\lambda} = \Big(\bigcup_{\gamma_{\widetilde{A}} \in u_{\widetilde{A}}} \{(\gamma_{\widetilde{A}})^{\lambda}\}, \bigcup_{\eta_{\widetilde{A}} \in v_{\widetilde{A}}} \{-(-\eta_{\widetilde{A}})^{\lambda}\} \Big); \tag{8.14}$$

where

$$\begin{aligned} 0 \le \gamma_{\widetilde{A}} \le 1, &\quad -1 \le \eta_{\widetilde{A}} \le 0, \\ 0 \le \gamma_{\widetilde{A}_1} \le 1, &\quad -1 \le \eta_{\widetilde{A}_1} \le 0, \\ 0 \le \gamma_{\widetilde{A}_2} \le 1, &\quad -1 \le \eta_{\widetilde{A}_2} \le 0. \end{aligned}$$

8.3 Hesitant Bipolar-Valued Neutrosophic Set

Awang et al [1] presented the concept of hesitant bipolar-valued neutrosophic set by combing the nations of bipolar neutrosophic set and hesitant fuzzy set. The presented notion indeed generalizes the notions of fuzzy set, intuitionistic fuzzy set, hesitant fuzzy set, single-valued neutrosophic set, single-valued neutrosophic hesitant fuzzy set, bipolar fuzzy set and bipolar neutrosophic set. Furthermore, Awang et al [1] gave a number of basic operational laws, union, intersection, complement and two aggregation operators for hesitant bipolar-valued neutrosophic sets.

Definition 8.3 ([1]) Let X be a reference set. A hesitant bipolar-valued neutrosophic set (HBVNS) $\mathbf{A}$ on X is defined in terms of six functions $u^-_{\mathbf{A}}(x)$, $u^+_{\mathbf{A}}(x)$, $w^-_{\mathbf{A}}(x)$, $w^+_{\mathbf{A}}(x)$, $v^-_{\mathbf{A}}(x)$, and $v^+_{\mathbf{A}}(x)$ as follows:

$$\mathbf{A} = \{\langle x, u^+_{\mathbf{A}}(x), w^+_{\mathbf{A}}(x), v^+_{\mathbf{A}}(x), u^-_{\mathbf{A}}(x), w^-_{\mathbf{A}}(x), v^-_{\mathbf{A}}(x), \rangle \mid x \in X\}, \quad (8.15)$$

where the positive hesitant bipolar-valued neutrosophic sets $u^+_{\mathbf{A}}(x), w^+_{\mathbf{A}}(x), v^+_{\mathbf{A}}(x) \in [0, 1]$ indicate, respectively, the possible satisfactory degree of truth, indeterminacy, and falsity of the element $x \in X$ corresponding to $\mathbf{A}$, and also, the negative hesitant bipolar-valued neutrosophic sets $u^-_{\mathbf{A}}(x), w^-_{\mathbf{A}}(x), v^-_{\mathbf{A}}(x) \in [-1, 0]$ indicate, respectively, the possible satisfactory degree of truth, indeterminacy, and falsity of the element $x \in X$ to the implicit counter property to the set $\mathbf{A}$. Furthermore, $0 \leq max\{u^+_{\mathbf{A}}(x)\} + \max\{w^+_{\mathbf{A}}(x)\} + \max\{v^+_{\mathbf{A}}(x)\} \leq 3$ and $-3 \leq max\{u^-_{\mathbf{A}}(x)\} + \max\{w^-_{\mathbf{A}}(x)\} + \max\{v^-_{\mathbf{A}}(x)\} \leq 0$ for any $x \in X$.

For the sake of simplicity, $h_{\mathbf{A}}(x) = (u^+_{\mathbf{A}}(x), w^+_{\mathbf{A}}(x), v^+_{\mathbf{A}}(x), u^-_{\mathbf{A}}(x), w^-_{\mathbf{A}}(x), v^-_{\mathbf{A}}(x))$ is called the hesitant bipolar-valued neutrosophic element (HBVNE).

Remark 8.3 Throughout this book, the set of all HBVNSs on the reference set X is denoted by $\mathbb{HBVNS}(X)$.

Example 8.4 Let $X = \{x_1, x_2\}$ be the reference set, $h_{\mathbf{A}}(x_1) = \{\langle\{0.2, 0.3\}, \{0.4\}, \{0.3\}, \{-0.1\}, \{-0.3\}, \{-0.6, -0.5\}\rangle\}$ and $h_{\mathbf{A}}(x_2) = \{\langle\{0.3\}, \{0.5\}, \{0.6, 0.7, 0.8\}, \{-0.1\}, \{-0.5\}, \{-0.2, -0.7, -0.9\}\rangle\}$ be the HBVNSs of x_i $(i = 1, 2)$ to a set $\mathbf{A}$, respectively. Then $\mathbf{A}$ can be considered as an HBVNS, i.e.,

$$\mathbf{A} = \{\langle x_1, \langle\{0.2, 0.3\}, \{0.4\}, \{0.3\}, \{-0.1\}, \{-0.3\}, \{-0.6, -0.5\}\rangle,$$

$$\langle x_2, \langle\{0.3\}, \{0.5\}, \{0.6, 0.7, 0.8\}, \{-0.1\}, \{-0.5\}, \{-0.2, -0.7, -0.9\}\rangle\}.$$

Given three HBVNEs represented by
$h_{\mathbf{A}} = \bigcup_{(\gamma_{\mathbf{A}}^{+},\delta_{\mathbf{A}}^{+},\eta_{\mathbf{A}}^{+},\gamma_{\mathbf{A}}^{-},\delta_{\mathbf{A}}^{-},\eta_{\mathbf{A}}^{-})\in(u_{\mathbf{A}}^{+}(x),w_{\mathbf{A}}^{+}(x),v_{\mathbf{A}}^{+}(x),u_{\mathbf{A}}^{-}(x),w_{\mathbf{A}}^{-}(x),v_{\mathbf{A}}^{-}(x))}$
$\{\langle \gamma_{\mathbf{A}}^{+}, \delta_{\mathbf{A}}^{+}, \eta_{\mathbf{A}}^{+}, \gamma_{\mathbf{A}}^{-}, \delta_{\mathbf{A}}^{-}, \eta_{\mathbf{A}}^{-}\rangle\}$,
$h_{\mathbf{A}_1} = \bigcup_{(\gamma_{\mathbf{A}_1}^{+},\delta_{\mathbf{A}_1}^{+},\eta_{\mathbf{A}_1}^{+},\gamma_{\mathbf{A}_1}^{-},\delta_{\mathbf{A}_1}^{-},\eta_{\mathbf{A}_1}^{-})\in(u_{\mathbf{A}_1}^{+}(x),w_{\mathbf{A}_1}^{+}(x),v_{\mathbf{A}_1}^{+}(x),u_{\mathbf{A}_1}^{-}(x),w_{\mathbf{A}_1}^{-}(x),v_{\mathbf{A}_1}^{-}(x))}$
$\{\langle \gamma_{\mathbf{A}_1}^{+}, \delta_{\mathbf{A}_1}^{+}, \eta_{\mathbf{A}_1}^{+}, \gamma_{\mathbf{A}_1}^{-}, \delta_{\mathbf{A}_1}^{-}, \eta_{\mathbf{A}_1}^{-}\rangle\}$ and
$h_{\mathbf{A}_2} = \bigcup_{(\gamma_{\mathbf{A}_2}^{+},\delta_{\mathbf{A}_2}^{+},\eta_{\mathbf{A}_2}^{+},\gamma_{\mathbf{A}_2}^{-},\delta_{\mathbf{A}_2}^{-},\eta_{\mathbf{A}_2}^{-})\in(u_{\mathbf{A}_2}^{+}(x),w_{\mathbf{A}_2}^{+}(x),v_{\mathbf{A}_2}^{+}(x),u_{\mathbf{A}_2}^{-}(x),w_{\mathbf{A}_2}^{-}(x),v_{\mathbf{A}_2}^{-}(x))}$
$\{\langle \gamma_{\mathbf{A}_2}^{+}, \delta_{\mathbf{A}_2}^{+}, \eta_{\mathbf{A}_2}^{+}, \gamma_{\mathbf{A}_2}^{-}, \delta_{\mathbf{A}_2}^{-}, \eta_{\mathbf{A}_2}^{-}\rangle\}$, some set and arithmetic operations on the HBVNEs, which are also HBVNEs, can be described as follows (see e.g. [1]):

$$(h_{\mathbf{A}})^c = \Big\{\Big\langle \bigcup_{\gamma_{\mathbf{A}}^{+}\in u_{\mathbf{A}}^{+}(x)} \{1-\gamma_{\mathbf{A}}^{+}\}, \bigcup_{\delta_{\mathbf{A}}^{+}\in w_{\mathbf{A}}^{+}(x)} \{1-\delta_{\mathbf{A}}^{+}\}, \bigcup_{\eta_{\mathbf{A}}^{+}\in v_{\mathbf{A}}^{+}(x)} \{1-\eta_{\mathbf{A}}^{+}\}, \bigcup_{\gamma_{\mathbf{A}}^{-}\in u_{\mathbf{A}}^{-}(x)} \{-1-\gamma_{\mathbf{A}}^{-}\}, \bigcup_{\delta_{\mathbf{A}}^{-}\in w_{\mathbf{A}}^{-}(x)} \{-1-\delta_{\mathbf{A}}^{-}\}, \bigcup_{\eta_{\mathbf{A}}^{-}\in v_{\mathbf{A}}^{-}(x)} \{-1-\eta_{\mathbf{A}}^{-}\}, \Big\rangle\Big\}; \tag{8.16}$$

$$h_{\mathbf{A}_1} \cup h_{\mathbf{A}_2} = \Big\{\Big\langle \bigcup_{\gamma_{\mathbf{A}_1}^{+}\in u_{\mathbf{A}_1}^{+}(x);\gamma_{\mathbf{A}_2}^{+}\in u_{\mathbf{A}_2}^{+}(x)} \max\{\gamma_{\mathbf{A}_1}^{+}, \gamma_{\mathbf{A}_2}^{+}\}, \bigcup_{\delta_{\mathbf{A}_1}^{+}\in w_{\mathbf{A}_1}^{+}(x);\delta_{\mathbf{A}_2}^{+}\in w_{\mathbf{A}_2}^{+}(x)} \Big\{\frac{\delta_{\mathbf{A}_1}^{+}+\delta_{\mathbf{A}_2}^{+}}{2}\Big\}, \bigcup_{\eta_{\mathbf{A}_1}^{+}\in v_{\mathbf{A}_1}^{+}(x);\eta_{\mathbf{A}_2}^{+}\in v_{\mathbf{A}_2}^{+}(x)} \min\{\eta_{\mathbf{A}_1}^{+}, \eta_{\mathbf{A}_2}^{+}\}, \bigcup_{\gamma_{\mathbf{A}_1}^{-}\in u_{\mathbf{A}_1}^{-}(x);\gamma_{\mathbf{A}_2}^{-}\in u_{\mathbf{A}_2}^{-}(x)} \min\{\gamma_{\mathbf{A}_1}^{-}, \gamma_{\mathbf{A}_2}^{-}\}, \bigcup_{\delta_{\mathbf{A}_1}^{-}\in w_{\mathbf{A}_1}^{-}(x);\delta_{\mathbf{A}_2}^{-}\in w_{\mathbf{A}_2}^{-}(x)} \{\frac{\delta_{\mathbf{A}_1}^{-}+\delta_{\mathbf{A}_2}^{-}}{2}\}, \bigcup_{\eta_{\mathbf{A}_1}^{-}\in v_{\mathbf{A}_1}^{-}(x);\eta_{\mathbf{A}_2}^{-}\in v_{\mathbf{A}_2}^{-}(x)} \max\{\eta_{\mathbf{A}_1}^{+}, \eta_{\mathbf{A}_2}^{+}\}, \Big\rangle\Big\}; \tag{8.17}$$

$$h_{\mathbf{A}_1} \cap h_{\mathbf{A}_2} = \Bigg\{\Bigg\langle \bigcup_{\gamma^+_{\mathbf{A}_1} \in u^+_{\mathbf{A}_1}(x); \gamma^+_{\mathbf{A}_2} \in u^+_{\mathbf{A}_2}(x)} \min\left\{\gamma^+_{\mathbf{A}_1}, \gamma^+_{\mathbf{A}_2}\right\},$$

$$\bigcup_{\delta^+_{\mathbf{A}_1} \in w^+_{\mathbf{A}_1}(x); \delta^+_{\mathbf{A}_2} \in w^+_{\mathbf{A}_2}(x)} \left\{\frac{\delta^+_{\mathbf{A}_1} + \delta^+_{\mathbf{A}_2}}{2}\right\},$$

$$\bigcup_{\eta^+_{\mathbf{A}_1} \in v^+_{\mathbf{A}_1}(x); \eta^+_{\mathbf{A}_2} \in v^+_{\mathbf{A}_2}(x)} \max\left\{\eta^+_{\mathbf{A}_1}, \eta^+_{\mathbf{A}_2}\right\},$$

$$\bigcup_{\gamma^-_{\mathbf{A}_1} \in u^-_{\mathbf{A}_1}(x); \gamma^-_{\mathbf{A}_2} \in u^-_{\mathbf{A}_2}(x)} \max\left\{\gamma^-_{\mathbf{A}_1}, \gamma^-_{\mathbf{A}_2}\right\},$$

$$\bigcup_{\delta^-_{\mathbf{A}_1} \in w^-_{\mathbf{A}_1}(x); \delta^-_{\mathbf{A}_2} \in w^-_{\mathbf{A}_2}(x)} \{\frac{\delta^-_{\mathbf{A}_1} + \delta^-_{\mathbf{A}_2}}{2}\},$$

$$\bigcup_{\eta^-_{\mathbf{A}_1} \in v^-_{\mathbf{A}_1}(x); \eta^-_{\mathbf{A}_2} \in v^-_{\mathbf{A}_2}(x)} \min\left\{\eta^+_{\mathbf{A}_1}, \eta^+_{\mathbf{A}_2}\right\}, \Bigg\rangle\Bigg\}; \tag{8.18}$$

$$h_{\mathbf{A}_1} \oplus h_{\mathbf{A}_2} = \Bigg\{\Bigg\langle \bigcup_{\gamma^+_{\mathbf{A}_1} \in u^+_{\mathbf{A}_1}(x); \gamma^+_{\mathbf{A}_2} \in u^+_{\mathbf{A}_2}(x)} \{\gamma^+_{\mathbf{A}_1} + \gamma^+_{\mathbf{A}_2} - \gamma^+_{\mathbf{A}_1} \times \gamma^+_{\mathbf{A}_2}\},$$

$$\bigcup_{\delta^+_{\mathbf{A}_1} \in w^+_{\mathbf{A}_1}(x); \delta^+_{\mathbf{A}_2} \in w^+_{\mathbf{A}_2}(x)} \{\delta^+_{\mathbf{A}_1} \times \delta^+_{\mathbf{A}_2}\}, \quad \bigcup_{\eta^+_{\mathbf{A}_1} \in v^+_{\mathbf{A}_1}(x); \eta^+_{\mathbf{A}_2} \in v^+_{\mathbf{A}_2}(x)} \{\eta^+_{\mathbf{A}_1} \times \eta^+_{\mathbf{A}_2}\},$$

$$\bigcup_{\gamma^-_{\mathbf{A}_1} \in u^-_{\mathbf{A}_1}(x); \gamma^-_{\mathbf{A}_2} \in u^-_{\mathbf{A}_2}(x)} \{-\gamma^-_{\mathbf{A}_1} \times \gamma^-_{\mathbf{A}_2}\},$$

$$\bigcup_{\delta^-_{\mathbf{A}_1} \in w^-_{\mathbf{A}_1}(x); \delta^-_{\mathbf{A}_2} \in w^-_{\mathbf{A}_2}(x)} \{-(-\delta^-_{\mathbf{A}_1} - \delta^-_{\mathbf{A}_2} - \delta^-_{\mathbf{A}_1} \times \delta^-_{\mathbf{A}_2})\},$$

$$\bigcup_{\eta^-_{\mathbf{A}_1} \in v^-_{\mathbf{A}_1}(x); \eta^-_{\mathbf{A}_2} \in v^-_{\mathbf{A}_2}(x)} \{-(-\eta^+_{\mathbf{A}_1} - \eta^+_{\mathbf{A}_2} - \eta^+_{\mathbf{A}_1} \times \eta^+_{\mathbf{A}_2})\}, \Bigg\rangle\Bigg\}; \tag{8.19}$$

$$h_{\mathbf{A}_1} \otimes h_{\mathbf{A}_2} = \Bigg\{\Bigg\langle \bigcup_{\gamma^+_{\mathbf{A}_1} \in u^+_{\mathbf{A}_1}(x); \gamma^+_{\mathbf{A}_2} \in u^+_{\mathbf{A}_2}(x)} \{\gamma^+_{\mathbf{A}_1} \times \gamma^+_{\mathbf{A}_2}\},$$

$$\bigcup_{\delta^+_{\mathbf{A}_1} \in w^+_{\mathbf{A}_1}(x); \delta^+_{\mathbf{A}_2} \in w^+_{\mathbf{A}_2}(x)} \{\delta^+_{\mathbf{A}_1} + \delta^+_{\mathbf{A}_2} - \delta^+_{\mathbf{A}_1} \times \delta^+_{\mathbf{A}_2}\},$$

$$\bigcup_{\eta^+_{\mathbf{A}_1}\in v^+_{\mathbf{A}_1}(x);\eta^+_{\mathbf{A}_2}\in v^+_{\mathbf{A}_2}(x)} \{\eta^+_{\mathbf{A}_1}+\eta^+_{\mathbf{A}_2}-\eta^+_{\mathbf{A}_1}\times\eta^+_{\mathbf{A}_2}\},$$

$$\bigcup_{\gamma^-_{\mathbf{A}_1}\in u^-_{\mathbf{A}_1}(x);\gamma^-_{\mathbf{A}_2}\in u^-_{\mathbf{A}_2}(x)} \{-(-\gamma^-_{\mathbf{A}_1}-\gamma^-_{\mathbf{A}_2}-\gamma^-_{\mathbf{A}_1}\times\gamma^-_{\mathbf{A}_2})\},$$

$$\bigcup_{\delta^-_{\mathbf{A}_1}\in w^-_{\mathbf{A}_1}(x);\delta^-_{\mathbf{A}_2}\in w^-_{\mathbf{A}_2}(x)} \{-\delta^-_{\mathbf{A}_1}\times\delta^-_{\mathbf{A}_2}\},$$

$$\bigcup_{\eta^-_{\mathbf{A}_1}\in v^-_{\mathbf{A}_1}(x);\eta^-_{\mathbf{A}_2}\in v^-_{\mathbf{A}_2}(x)} \{-\eta^+_{\mathbf{A}_1}\times\eta^+_{\mathbf{A}_2}\},\Big\rangle\Big\}; \tag{8.20}$$

$$\lambda h_{\mathbf{A}} = \Big\{\Big\langle \bigcup_{\gamma^+_{\mathbf{A}}\in u^+_{\mathbf{A}}(x)} \{1-(1-\gamma^+_{\mathbf{A}})^\lambda\},$$

$$\bigcup_{\delta^+_{\mathbf{A}}\in w^+_{\mathbf{A}}(x)} \{(\delta^+_{\mathbf{A}})^\lambda\},$$

$$\bigcup_{\eta^+_{\mathbf{A}}\in v^+_{\mathbf{A}}(x)} \{(\eta^+_{\mathbf{A}})^\lambda\},$$

$$\bigcup_{\gamma^-_{\mathbf{A}}\in u^-_{\mathbf{A}}(x)} \{-(-\gamma^-_{\mathbf{A}})^\lambda\},$$

$$\bigcup_{\delta^-_{\mathbf{A}}\in w^-_{\mathbf{A}}(x)} \{-(1-(1-(-\delta^-_{\mathbf{A}}))^\lambda)\},$$

$$\bigcup_{\eta^-_{\mathbf{A}}\in v^-_{\mathbf{A}}(x)} \{-(1-(1-\eta^-_{\mathbf{A}}))^\lambda)\},\Big\rangle\Big\}; \tag{8.21}$$

$$(h_{\mathbf{A}})^\lambda = \Big\{\Big\langle \bigcup_{\gamma^+_{\mathbf{A}}\in u^+_{\mathbf{A}}(x)} \{(\gamma^+_{\mathbf{A}})^\lambda\},$$

$$\bigcup_{\delta^+_{\mathbf{A}}\in w^+_{\mathbf{A}}(x)} \{1-(1-\delta^+_{\mathbf{A}})^\lambda\},$$

$$\bigcup_{\eta^+_{\mathbf{A}}\in v^+_{\mathbf{A}}(x)} \{1-(1-\eta^+_{\mathbf{A}})^\lambda\},$$

$$\bigcup_{\gamma^-_{\mathbf{A}}\in u^-_{\mathbf{A}}(x)} \{-(1-(1-(-\gamma^-_{\mathbf{A}}))^\lambda)\},$$

$$\bigcup_{\delta_{\mathbf{A}}^{-}\in w_{\mathbf{A}}^{-}(x)} \{-(-\delta_{\mathbf{A}}^{-})^{\lambda}\},$$

$$\bigcup_{\eta_{\mathbf{A}}^{-}\in \upsilon_{\mathbf{A}}^{-}(x)} \{-(-\eta_{\mathbf{A}}^{-})^{\lambda}\},\Big\rangle\Big\}; \quad \lambda > 0. \tag{8.22}$$

References

1. A. Awang, M. Ali, L. Abdullah, Hesitant bipolar-valued neutrosophic set: formulation, theory and application. IEEE Access **7**, 176099–176114 (2019)
2. H. Gao, M. Lu, Y. Wei, Dual hesitant bipolar fuzzy hamacher aggregation operators and their applications to multiple attribute decision making. J. Intell. Fuzzy Syst. **37**, 5755–5766 (2019)
3. Z. Gul, Some bipolar fuzzy aggregations operators and their applications in multicriteria group decision making, M. Phil Thesis, 2015.
4. Y. Han, Q. Luo, Chen, Hesitant bipolar fuzzy set and its application in decision making, in *Fuzzy Systems and Data Mining II: Proceedings of FSDM 2016*, ed. by S.L. Sun, A.J. Tallon-Ballesteros, D.S. Pamucar (2016)
5. G. Wei, F.E. Alsaadi, T. Hayat, A. Alsaedi, Hesitant bipolar fuzzy aggregation operators in multiple attribute decision making. J. Intell. Fuzzy Syst. **33**, 1119–1128 (2017)
6. X.R. Xu, G.W. Wei, Dual hesitant bipolar fuzzy aggregation operators in multiple attribute decision making. Int. J. Knowl. Based Intell. Eng. Syst. **21**, 155–164 (2017)
7. W.R. Zhang, Bipolar fuzzy sets and relations: A computational frame work for cognitive modelling and multiagent decision analysis, in *NAFIPS/IFIS/NASA '94. Proceedings of the First International Joint Conference of The North American Fuzzy Information Processing Society Biannual Conference. The Industrial Fuzzy Control and Intellige* (1994) pp. 305–309

Chapter 9
Cubic Hesitant Fuzzy Set

Abstract In this chapter, we first introduce the concept of cubic hesitant fuzzy set. Then, we present the form that in which cubic hesitant fuzzy set is characterized in the form of triangular, and it is known as triangular cubic hesitant fuzzy set. In the next part of this chapter, we deal with the definition of linguistic intuitionistic cubic hesitant variable.

9.1 Cubic Hesitant Fuzzy Set

Fu et al. [1] expressed the hybrid information of both cubic fuzzy set and hesitant fuzzy set, and called it cubic hesitant fuzzy set. They then studied the generalized form of distance and similarity measures of cubic hesitant fuzzy sets. Yong et al. [8] developed a Jaccard similarity measure for cubic hesitant fuzzy sets and implemented it in a multiple criteria decision making technique under cubic hesitant environment. Mahmood et al. [6] investigated and discussed about the concepts of internal (or external) cubic hesitant fuzzy set, P(or R)-union, P(or R)-intersection of cubic hesitant fuzzy sets, P(or R)-addition, and P(or R)-multiplication of cubic hesitant fuzzy sets.

The concept of cubic fuzzy set A on X is defined as (see [5])

$$A = \{\langle x, \iota_A(x), \gamma_A(x)\rangle \mid x \in X\}, \tag{9.1}$$

in which $\iota_A(x) = [\iota_A^L, \iota_A^U]$ is an interval-valued fuzzy number and $\gamma_A(x)$ is a unique fuzzy value for the element $x \in X$.

B. Farhadinia, *Hesitant Fuzzy Set*, Computational Intelligence Methods and Applications, https://doi.org/10.1007/978-981-16-7301-6_9

Suppose that for any $x \in X$, $A_1 = (\iota_{A_1}(x), \gamma_{A_1}(x))$ and $A_2 = (\iota_{A_2}(x), \gamma_{A_2}(x))$ are to be two cubic fuzzy sets. Then,

1. (Equality): $A_1 = A_2$ if and only if $\iota_{A_1}(x) = \iota_{A_2}(x)$ and $\gamma_{A_1}(x) = \gamma_{A_2}(x)$;
2. (P-order): $A_1 \subseteq_P A_2$ if and only if $\iota_{A_1}(x) \subseteq \iota_{A_2}(x)$ and $\gamma_{A_1}(x) \leq \gamma_{A_2}(x)$;
3. (R-order): $A_1 \subseteq_R A_2$ if and only if $\iota_{A_1}(x) \subseteq \iota_{A_2}(x)$ and $\gamma_{A_1}(x) \geq \gamma_{A_2}(x)$.

Definition 9.1 ([6]) Let X be a reference set. A cubic hesitant fuzzy set (CHFS) A on X is defined in terms of $u_A(x)$ and $v_A(x)$ as follows:

$$\begin{aligned} A &= \{\langle x, u_A(x), v_A(x)\rangle \mid x \in X\} \\ &= \left\{\left\langle x, \bigcup_{[\gamma_A^L, \gamma_A^U] \in u_A(x)} \left\{[\gamma_A^L, \gamma_A^U]\right\}, \bigcup_{\eta_A \in v_A(x)} \{\eta_A\}\right\rangle \mid x \in X\right\}, \end{aligned} \tag{9.2}$$

where $u_A(x)$ is an interval-valued hesitant fuzzy element, and $v_A(x)$ is the set of some different values in [0, 1] arranging in an increasing order for the element $x \in X$ to A.

For the sake of simplicity, $h_A(x) = \langle u_A(x), v_A(x)\rangle$ is called the cubic hesitant fuzzy number (CHFN).

Remark 9.1 Throughout this book, the set of all CHFSs on the reference set X is denoted by $\mathbb{CHFS}(X)$.

Example 9.1 Let $X = \{x_1, x_2\}$ be the reference set, $h_A(x_1) = \langle u_A(x_1), v_A(x_1)\rangle = \langle\{[0.1, 0.3], [0.2, 0.5]\}, \{0.3, 0.4, 0.6\}\rangle$ and $h_A(x_2) = \langle u_A(x_2), v_A(x_2)\rangle = \langle\{[0.4, 0.8]\}, \{0.1, 0.4, 0.5, 0.9\}\rangle$ be the CHFEs of x_i $(i = 1, 2)$ to a set A, respectively. Then A can be considered as a CHFS, i.e.,

$$\begin{aligned} A = \{&\langle x_1, \langle\{[0.1, 0.3], [0.2, 0.5]\}, \{0.3, 0.4, 0.6\}\rangle\rangle, \\ &\langle x_2, \langle\{[0.4, 0.8]\}, \{0.1, 0.4, 0.5, 0.9\}\rangle\rangle\}. \end{aligned}$$

Suppose that for any $x \in X$, $h_{A_1}(x) = \langle u_{A_1}(x), v_{A_1}(x)\rangle$ and $h_{A_2}(x) = \langle u_{A_2}(x), v_{A_2}(x)\rangle$ are to be two CHFEs. Then (see [6])

1. (Equality): $h_{A_1} = h_{A_2}$ if and only if $u_{A_1}(x) = u_{A_2}(x)$ and $v_{A_1}(x) = v_{A_2}(x)$;
2. (P-order): $h_{A_1} \subseteq_P h_{A_2}$ if and only if $u_{A_1}(x) \subseteq u_{A_2}(x)$ and $v_{A_1}(x) \leq v_{A_2}(x)$;
3. (R-order): $h_{A_1} \subseteq_R h_{A_2}$ if and only if $u_{A_1}(x) \subseteq u_{A_2}(x)$ and $v_{A_1}(x) \geq v_{A_2}(x)$.

Given a fixed $x \in X$, some operational laws on CHFNs are defined as follows [6]:

$$A_1{}^c = \{\langle x, 1 - u_{A_1}(x), 1 - v_{A_1}(x)\rangle \mid x \in X\};$$

$$A_1 \cup_P A_2 = \{\langle x, \max\{u_{A_1}(x), u_{A_2}(x)\}, \max\{v_{A_1}(x), v_{A_2}(x)\}\rangle \mid x \in X\};$$

$$A_1 \cap_P A_2 = \{\langle x, \min\{u_{A_1}(x), u_{A_2}(x)\}, \min\{v_{A_1}(x), v_{A_2}(x)\}\rangle \mid x \in X\};$$

$$A_1 \cup_R A_2 = \{\langle x, \max\{u_{A_1}(x), u_{A_2}(x)\}, \min\{v_{A_1}(x), v_{A_2}(x)\}\rangle \mid x \in X\};$$

$$A_1 \cap_R A_2 = \{\langle x, \min\{u_{A_1}(x), u_{A_2}(x)\}, \max\{v_{A_1}(x), v_{A_2}(x)\}\rangle \mid x \in X\};$$

$$A_1 \oplus_P A_2 = \{\langle x, u_{A_1}(x) \oplus u_{A_2}(x), v_{A_1}(x) \oplus v_{A_2}(x)\rangle \mid x \in X\};$$

$$A_1 \otimes_P A_2 = \{\langle x, u_{A_1}(x) \otimes u_{A_2}(x), v_{A_1}(x) \otimes v_{A_2}(x)\rangle \mid x \in X\};$$

$$A_1 \oplus_R A_2 = \{\langle x, u_{A_1}(x) \oplus u_{A_2}(x), v_{A_1}(x) \otimes v_{A_2}(x)\rangle \mid x \in X\};$$

$$A_1 \otimes_R A_2 = \{\langle x, u_{A_1}(x) \otimes u_{A_2}(x), v_{A_1}(x) \oplus v_{A_2}(x)\rangle \mid x \in X\};$$

where

$$1 - u_{A_1}(x) = \bigcup_{[\gamma^L_{A_1}, \gamma^U_{A_1}] \in u_{A_1}(x)} \{[1 - \gamma^U_{A_1}, 1 - \gamma^L_{A_1}]\};$$

$$1 - v_{A_1}(x) = \bigcup_{\eta_{A_1} \in v_{A_1}(x)} \{1 - \eta_{A_1}\};$$

$$\begin{aligned}
\max\{u_{A_1}(x), u_{A_2}(x)\} &= \max\left\{ \bigcup_{\left[\gamma^L_{A_1}, \gamma^U_{A_1}\right] \in u_{A_1}(x)} \left\{\left[\gamma^L_{A_1}, \gamma^U_{A_1}\right]\right\}, \bigcup_{\left[\gamma^L_{A_2}, \gamma^U_{A_2}\right] \in u_{A_2}(x)} \left\{\left[\gamma^L_{A_2}, \gamma^U_{A_2}]\right]\right\} \right. \\
&= \bigcup_{\left[\gamma^L_{A_1}, \gamma^U_{A_1}\right] \in u_{A_1}(x); \left[\gamma^L_{A_2}, \gamma^U_{A_2}\right] \in u_{A_2}(x)} \left\{\left[\max\left\{\gamma^L_{A_1}, \gamma^L_{A_2}\right\}, \max\left\{\gamma^U_{A_1}, \gamma^U_{A_2}\right\}\right]\right\},
\end{aligned}$$

$$\begin{aligned}
\max\{v_{A_1}(x), v_{A_2}(x)\} &= \max\left\{ \bigcup_{\eta_{A_1} \in v_{A_1}(x)} \{\eta_{A_1}\}, \bigcup_{\eta_{A_2} \in v_{A_2}(x)} \{\eta_{A_2}\} \right\} \\
&= \bigcup_{\eta_{A_1} \in v_{A_1}(x); \eta_{A_2} \in v_{A_2}(x)} \{\max\{\eta_{A_1}, \eta_{A_2}\}\};
\end{aligned}$$

$$\begin{aligned}
\min\{u_{A_1}(x), u_{A_2}(x)\} &= \min\left\{ \bigcup_{\left[\gamma^L_{A_1}, \gamma^U_{A_1}\right] \in u_{A_1}(x)} \left\{\left[\gamma^L_{A_1}, \gamma^U_{A_1}\right]\right\}, \bigcup_{\left[\gamma^L_{A_2}, \gamma^U_{A_2}\right] \in u_{A_2}(x)} \left\{\left[\gamma^L_{A_2}, \gamma^U_{A_2}\right]\right\} \right\} \\
&= \bigcup_{\left[\gamma^L_{A_1}, \gamma^U_{A_1}\right] \in u_{A_1}(x); \left[\gamma^L_{A_2}, \gamma^U_{A_2}\right] \in u_{A_2}(x)} \left\{\left[\min\{\gamma^L_{A_1}, \gamma^L_{A_2}\}, \min\left\{\gamma^U_{A_1}, \gamma^U_{A_2}\right\}\right]\right\},
\end{aligned}$$

$$\begin{aligned}
\min\{v_{A_1}(x), v_{A_2}(x)\} &= \min\left\{ \bigcup_{\eta_{A_1} \in v_{A_1}(x)} \{\eta_{A_1}\}, \bigcup_{\eta_{A_2} \in v_{A_2}(x)} \{\eta_{A_2}\} \right\} \\
&= \bigcup_{\eta_{A_1} \in v_{A_1}(x); \eta_{A_2} \in v_{A_2}(x)} \{\min\{\eta_{A_1}, \eta_{A_2}\}\};
\end{aligned}$$

$$u_{A_1}(x) \oplus u_{A_2}(x) = \bigcup_{\left[\gamma_{A_1}^L, \gamma_{A_1}^U\right] \in u_{A_1}(x); \left[\gamma_{A_2}^L, \gamma_{A_2}^U\right] \in u_{A_2}(x)} \times \left\{\left[\gamma_{A_1}^L + \gamma_{A_2}^L - \gamma_{A_1}^L \times \gamma_{A_2}^L, \gamma_{A_1}^U + \gamma_{A_2}^U - \gamma_{A_1}^U \times \gamma_{A_2}^U\right]\right\},$$

$$u_{A_1}(x) \otimes u_{A_2}(x) = \bigcup_{\left[\gamma_{A_1}^L, \gamma_{A_1}^U\right] \in u_{A_1}(x); \left[\gamma_{A_2}^L, \gamma_{A_2}^U\right] \in u_{A_2}(x)} \left\{\left[\gamma_{A_1}^L \times \gamma_{A_2}^L, \gamma_{A_1}^U \times \gamma_{A_2}^U\right]\right\},$$

$$v_{A_1}(x) \oplus v_{A_2}(x) = \bigcup_{\eta_{A_1} \in v_{A_1}(x); \eta_{A_2} \in v_{A_2}(x)} \{\eta_{A_1} + \eta_{A_2} - \eta_{A_1} \times \eta_{A_2}\};$$

$$v_{A_1}(x) \otimes v_{A_2}(x) v = \bigcup_{\eta_{A_1} \in v_{A_1}(x); \eta_{A_2} \in v_{A_2}(x)} \{\eta_{A_1} \times \eta_{A_2}\}.$$

9.2 Triangular Cubic Hesitant Fuzzy Set

Fahmi et al. [1] presented the concept of triangular cubic hesitant fuzzy set, and further some triangular cubic hesitant fuzzy aggregation operators. Fahmi et al. [2] developed a set of basic operational laws for triangular cubic hesitant fuzzy sets, and on the basis of triangular cubic hesitant fuzzy hamming distance, they represented a TOPSIS technique under triangular cubic hesitant fuzzy environment,

Definition 9.2 ([1]) Let X be a reference set. A triangular cubic hesitant fuzzy set (TCHFS) $\widetilde{A}$ on X is defined in terms of $u_{\widetilde{A}}(x)$ and $v_{\widetilde{A}}(x)$ as follows:

$$\widetilde{A} = \{\langle x, u_{\widetilde{A}}(x), v_{\widetilde{A}}(x)\rangle \mid x \in X\}, \tag{9.3}$$

where

$$u_{\widetilde{A}}(x) = \bigcup_{\left[\left(\gamma_{1\widetilde{A}}^L, \gamma_{2\widetilde{A}}^L, \gamma_{3\widetilde{A}}^L\right), \left(\gamma_{1\widetilde{A}}^U, \gamma_{2\widetilde{A}}^U, \gamma_{3\widetilde{A}}^U\right)\right] \in u_{\widetilde{A}}(x)} \left\{\left[\left(\gamma_{1\widetilde{A}}^L, \gamma_{2\widetilde{A}}^L, \gamma_{3\widetilde{A}}^L\right), \left(\gamma_{1\widetilde{A}}^U, \gamma_{2\widetilde{A}}^U, \gamma_{3\widetilde{A}}^U\right)\right]\right\},$$

$$v_{\widetilde{A}}(x) = \bigcup_{(\eta_{1\widetilde{A}}, \eta_{2\widetilde{A}}, \eta_{3\widetilde{A}} \in v_{\widetilde{A}}(x))} \{(\eta_{1\widetilde{A}}, \eta_{2\widetilde{A}}, \eta_{3\widetilde{A}})\}.$$

Here, $u_{\widetilde{A}}(x)$ is an interval-valued triangular hesitant fuzzy set, and $v_{\widetilde{A}}(x)$ is the set of some different triangular hesitant fuzzy values in $[0, 1]$ for the element $x \in X$ to $\widetilde{A}$.

For the sake of simplicity, $h_{\widetilde{A}}(x) = \langle u_{\widetilde{A}}(x), v_{\widetilde{A}}(x)\rangle$ is called the triangular cubic hesitant fuzzy number (TCHFN).

Remark 9.2 Throughout this book, the set of all TCHFSs on the reference set X is denoted by $\mathbb{TCHFS}(X)$.

Example 9.2 Let $X = \{x_1, x_2\}$ be the reference set, $h_{\widetilde{A}}(x_1) = \langle u_{\widetilde{A}}(x_1), v_{\widetilde{A}}(x_1)\rangle = \langle\{[(0.1, 0.3, 0.5), (0.3, 0.4, 0.6)]\}, \{(0.2, 0.5, 0.9)\}\rangle$ and $h_{\widetilde{A}}(x_2) = \langle u_{\widetilde{A}}(x_2), v_{\widetilde{A}}(x_2)\rangle = \langle\{[(0.1, 0.2, 0.3), (0.1, 0.4, 0.6)]\}, \{(0.1, 0.5, 0.8)\}\rangle$ be the TCHFEs of x_i $(i = 1, 2)$ to a set $\widetilde{A}$, respectively. Then $\widetilde{A}$ can be considered as an TCHFS, i.e.,

$$\begin{aligned}\widetilde{A} = \{&\langle x_1, \langle\{[(0.1, 0.3, 0.5), (0.3, 0.4, 0.6)]\}, \{(0.2, 0.5, 0.9)\}\rangle\rangle,\\ &\langle x_2, \langle\{[(0.1, 0.2, 0.3), (0.1, 0.4, 0.6)]\}, \{(0.1, 0.5, 0.8)\}\rangle\rangle\}.\end{aligned}$$

Given a fixed $x \in X$, some operational laws on the TCHFNs $h_{\widetilde{A}} = \langle u_{\widetilde{A}}(x), v_{\widetilde{A}}(x)\rangle = \langle\bigcup_{[(\gamma^L_{1\widetilde{A}},\gamma^L_{2\widetilde{A}},\gamma^L_{3\widetilde{A}}),(\gamma^U_{1\widetilde{A}},\gamma^U_{2\widetilde{A}},\gamma^U_{3\widetilde{A}})]\in u_{\widetilde{A}}(x)}\{[(\gamma^L_{1\widetilde{A}}, \gamma^L_{2\widetilde{A}}, \gamma^L_{3\widetilde{A}}), (\gamma^U_{1\widetilde{A}}, \gamma^U_{2\widetilde{A}}, \gamma^U_{3\widetilde{A}})]\}, \bigcup_{(\eta_{1\widetilde{A}},\eta_{2\widetilde{A}},\eta_{3\widetilde{A}}\in v_{\widetilde{A}}(x)}\{(\eta_{1\widetilde{A}}, \eta_{2\widetilde{A}}, \eta_{3\widetilde{A}})\}\rangle$,
$h_{\widetilde{A}_1} = \langle u_{\widetilde{A}_1}(x), v_{\widetilde{A}_1}(x)\rangle$
$=\langle\bigcup_{[(\gamma^L_{1\widetilde{A}_1},\gamma^L_{2\widetilde{A}_1},\gamma^L_{3\widetilde{A}_1}),(\gamma^U_{1\widetilde{A}_1},\gamma^U_{2\widetilde{A}_1},\gamma^U_{3\widetilde{A}_1})]\in u_{\widetilde{A}_1}(x)}\{[(\gamma^L_{1\widetilde{A}_1}, \gamma^L_{2\widetilde{A}_1}, \gamma^L_{3\widetilde{A}_1}), (\gamma^U_{1\widetilde{A}_1}, \gamma^U_{2\widetilde{A}_1}, \gamma^U_{3\widetilde{A}_1})]\}$
$, \bigcup_{(\eta_{1\widetilde{A}_1},\eta_{2\widetilde{A}_1},\eta_{3\widetilde{A}_1}\in v_{\widetilde{A}_1}(x)}\{(\eta_{1\widetilde{A}_1}, \eta_{2\widetilde{A}_1}, \eta_{3\widetilde{A}_1})\}\rangle$ and
$h_{\widetilde{A}_2} = \langle u_{\widetilde{A}_2}(x), v_{\widetilde{A}_2}(x)\rangle$
$=\langle\bigcup_{[(\gamma^L_{1\widetilde{A}_2},\gamma^L_{2\widetilde{A}_2},\gamma^L_{3\widetilde{A}_2}),(\gamma^U_{1\widetilde{A}_2},\gamma^U_{2\widetilde{A}_2},\gamma^U_{3\widetilde{A}_2})]\in u_{\widetilde{A}_2}(x)}\{[(\gamma^L_{1\widetilde{A}_2}, \gamma^L_{2\widetilde{A}_2}, \gamma^L_{3\widetilde{A}_2}), (\gamma^U_{1\widetilde{A}_2}, \gamma^U_{2\widetilde{A}_2}, \gamma^U_{3\widetilde{A}_2})]\}$
$, \bigcup_{(\eta_{1\widetilde{A}_2},\eta_{2\widetilde{A}_2},\eta_{3\widetilde{A}_2}\in v_{\widetilde{A}_2}(x)}\{(\eta_{1\widetilde{A}_2}, \eta_{2\widetilde{A}_2}, \eta_{3\widetilde{A}_2})\}\rangle$ are defined as the followings based on the Einstein t-norm $T(x, y) = \frac{xy}{1+(1-x)(1-y)}$ with its dual t-conorm $S(x, y) = \frac{x+y}{1+xy}$ [1]:

$$h_{\widetilde{A}_1} \oplus h_{\widetilde{A}_2} =$$

$$\Bigg\langle \bigcup_{\left[\left(\gamma^L_{1\widetilde{A}_i},\gamma^L_{2\widetilde{A}_i},\gamma^L_{3\widetilde{A}_i}\right),\left(\gamma^U_{1\widetilde{A}_i},\gamma^U_{2\widetilde{A}_i},\gamma^U_{3\widetilde{A}_i}\right)\right]\in u_{\widetilde{A}_i}(x),\ i=1,2} \Bigg\{\Bigg[\Bigg(\frac{\gamma^L_{1\widetilde{A}_1} + \gamma^L_{1\widetilde{A}_2}}{1 + \gamma^L_{1\widetilde{A}_1} + \gamma^L_{1\widetilde{A}_2}},$$

$$\frac{\gamma^L_{2\widetilde{A}_1} + \gamma^L_{2\widetilde{A}_2}}{1 + \gamma^L_{2\widetilde{A}_1} + \gamma^L_{2\widetilde{A}_2}},$$

$$\frac{\gamma^L_{3\widetilde{A}_1} + \gamma^L_{3\widetilde{A}_2}}{1 + \gamma^L_{3\widetilde{A}_1} + \gamma^L_{3\widetilde{A}_2}}\Bigg),$$

$$\Bigg(\frac{\gamma^U_{1\widetilde{A}_1} + \gamma^U_{1\widetilde{A}_2}}{1 + \gamma^U_{1\widetilde{A}_1} + \gamma^U_{1\widetilde{A}_2}},$$

$$\frac{\gamma^U_{2\widetilde{A}_1}+\gamma^U_{2\widetilde{A}_2}}{1+\gamma^U_{2\widetilde{A}_1}+\gamma^U_{2\widetilde{A}_2}},$$

$$\frac{\gamma^U_{3\widetilde{A}_1}+\gamma^U_{3\widetilde{A}_2}}{1+\gamma^U_{3\widetilde{A}_1}+\gamma^U_{3\widetilde{A}_2}}),]\},$$

$$\bigcup_{(\eta_{1\widetilde{A}_i},\eta_{2\widetilde{A}_i},\eta_{3\widetilde{A}_i})\in v_{\widetilde{A}_i}(x),\ i=1,2}\{(\frac{\eta_{1\widetilde{A}_1}\times\delta_{1\widetilde{A}_2}}{(1+(1-\delta_{1\widetilde{A}_1}))(1-\delta_{1\widetilde{A}_2})},$$

$$\frac{\eta_{2\widetilde{A}_1}\times\delta_{2\widetilde{A}_2}}{(1+(1-\delta_{2\widetilde{A}_1}))(1-\delta_{2\widetilde{A}_2})},$$

$$\frac{\eta_{3\widetilde{A}_1}\times\delta_{3\widetilde{A}_2}}{(1+(1-\delta_{3\widetilde{A}_1}))(1-\delta_{3\widetilde{A}_2})})\}\rangle; \tag{9.4}$$

$$\lambda h_{\widetilde{A}}=$$

$$\langle\bigcup_{[(\gamma^L_{1\widetilde{A}},\gamma^L_{2\widetilde{A}},\gamma^L_{3\widetilde{A}}),(\gamma^U_{1\widetilde{A}},\gamma^U_{2\widetilde{A}},\gamma^U_{3\widetilde{A}})]\in u_{\widetilde{A}}(x)}\{[(\frac{(1+\gamma^L_{1\widetilde{A}_1})^\lambda-(1-\gamma^L_{1\widetilde{A}_1})^\lambda}{(1+\gamma^L_{1\widetilde{A}_1})^\lambda+(1-\gamma^L_{1\widetilde{A}_1})^\lambda},$$

$$\frac{(1+\gamma^L_{2\widetilde{A}_1})^\lambda-(1-\gamma^L_{2\widetilde{A}_1})^\lambda}{(1+\gamma^L_{2\widetilde{A}_1})^\lambda+(1-\gamma^L_{2\widetilde{A}_1})^\lambda},$$

$$\frac{(1+\gamma^L_{3\widetilde{A}_1})^\lambda-(1-\gamma^L_{3\widetilde{A}_1})^\lambda}{(1+\gamma^L_{3\widetilde{A}_1})^\lambda+(1-\gamma^L_{3\widetilde{A}_1})^\lambda}),$$

$$(\frac{(1+\gamma^U_{1\widetilde{A}_1})^\lambda-(1-\gamma^U_{1\widetilde{A}_1})^\lambda}{(1+\gamma^U_{1\widetilde{A}_1})^\lambda+(1-\gamma^U_{1\widetilde{A}_1})^\lambda},$$

$$\frac{(1+\gamma^U_{2\widetilde{A}_1})^\lambda-(1-\gamma^U_{2\widetilde{A}_1})^\lambda}{(1+\gamma^U_{2\widetilde{A}_1})^\lambda+(1-\gamma^U_{2\widetilde{A}_1})^\lambda},$$

$$\frac{(1+\gamma^U_{3\widetilde{A}_1})^\lambda-(1-\gamma^U_{3\widetilde{A}_1})^\lambda}{(1+\gamma^U_{3\widetilde{A}_1})^\lambda+(1-\gamma^U_{3\widetilde{A}_1})^\lambda})]\},$$

$$\bigcup_{(\eta_{1\widetilde{A}},\eta_{2\widetilde{A}},\eta_{3\widetilde{A}})\in v_{\widetilde{A}}(x)}\{(\frac{2(\delta_{1\widetilde{A}})^\lambda}{(2-\delta_{1\widetilde{A}})^\lambda+(\delta_{1\widetilde{A}})^\lambda},$$

$$\frac{2(\delta_{2\widetilde{A}})^{\lambda}}{(2-\delta_{2\widetilde{A}})^{\lambda}+(\delta_{2\widetilde{A}})^{\lambda}},$$

$$\left.\left.\frac{2(\delta_{3\widetilde{A}})^{\lambda}}{(2-\delta_{3\widetilde{A}})^{\lambda}+(\delta_{3\widetilde{A}})^{\lambda}},\right)\right\}\right\rangle. \tag{9.5}$$

9.3 Linguistic Intuitionistic Cubic Hesitant Variable

Qiyas et al. [7] introduced the concept of linguistic intuitionistic cubic hesitant variable, and furthermore, they presented a number of aggregation operators to design a model for solving a multiple criteria decision making issues under linguistic intuitionistic cubic hesitant environment.

We consider $\mathfrak{S} = \{s_\alpha \mid \alpha \in [0, \tau]\}$ as a continuous linguistic term set. Then, a linguistic intuitionistic cubic hesitant variable is defined as the following:

Definition 9.3 ([7]) Let X be a reference set. A linguistic intuitionistic cubic hesitant variable (LICHV) $\mathbf{A}^{\mathfrak{S}}$ on X is defined in terms of $u_{\mathbf{A}^{\mathfrak{S}}}(x)$ and $v_{\mathbf{A}^{\mathfrak{S}}}(x)$ as follows:

$$\mathbf{A}^{\mathfrak{S}} = \{\langle x, u_{\mathbf{A}^{\mathfrak{S}}}(x), v_{\mathbf{A}^{\mathfrak{S}}}(x)\rangle \mid x \in X\}, \tag{9.6}$$

where

$$u_{\mathbf{A}^{\mathfrak{S}}}(x) = \left\langle \left[s^L_{\gamma_{\mathbf{A}^{\mathfrak{S}}}}, s^U_{\gamma_{\mathbf{A}^{\mathfrak{S}}}}\right], \bigcup_{s_{\gamma_{\mathbf{A}^{\mathfrak{S}}}} \in u_{\mathbf{A}^{\mathfrak{S}}}(x)} \{s_{\gamma_{\mathbf{A}^{\mathfrak{S}}}}\}\right\rangle,$$

$$v_{\mathbf{A}^{\mathfrak{S}}}(x) = \left\langle \left[s^L_{\eta_{\mathbf{A}^{\mathfrak{S}}}}, s^U_{\eta_{\mathbf{A}^{\mathfrak{S}}}}\right], \bigcup_{s_{\eta_{\mathbf{A}^{\mathfrak{S}}}} \in v_{\mathbf{A}^{\mathfrak{S}}}(x)} \{s_{\eta_{\mathbf{A}^{\mathfrak{S}}}}\}\right\rangle,$$

where $[s^L_{\gamma_{\mathbf{A}^{\mathfrak{S}}}}, s^U_{\gamma_{\mathbf{A}^{\mathfrak{S}}}}]$ (or $[s^L_{\eta_{\mathbf{A}^{\mathfrak{S}}}}, s^U_{\eta_{\mathbf{A}^{\mathfrak{S}}}}]$) is an interval-valued element, and $\bigcup_{s_{\gamma_{\mathbf{A}^{\mathfrak{S}}}} \in u_{\mathbf{A}^{\mathfrak{S}}}(x)}\{s_{\gamma_{\mathbf{A}^{\mathfrak{S}}}}\}$ (or $\bigcup_{s_{\eta_{\mathbf{A}^{\mathfrak{S}}}} \in v_{\mathbf{A}^{\mathfrak{S}}}(x)}\{s_{\eta_{\mathbf{A}^{\mathfrak{S}}}}\}$) is the set of some different values in [0, 1] arranging in an increasing order for the element $x \in X$ to $\mathbf{A}^{\mathfrak{S}}$.

For the sake of simplicity, $h_{\mathbf{A}^{\mathfrak{S}}}(x) = (u_{\mathbf{A}^{\mathfrak{S}}}(x), v_{\mathbf{A}^{\mathfrak{S}}}(x))$ is called the linguistic intuitionistic cubic hesitant variable (LICHV).

Remark 9.3 Throughout this book, the set of all LICHVs on the reference set X is denoted by $\mathbb{LICHV}(X)$.

Generally, if two LICHVs $h_{\mathbf{A}^{\mathfrak{S}}}(x_1)$ and $h_{\mathbf{A}^{\mathfrak{S}}}(x_2)$ have different numbers in their HFS parts of $u_{\mathbf{A}^{\mathfrak{S}}}(x_1)$, $v_{\mathbf{A}^{\mathfrak{S}}}(x_1)$, $u_{\mathbf{A}^{\mathfrak{S}}}(x_2)$ and $v_{\mathbf{A}^{\mathfrak{S}}}(x_2)$, then the HFS parts should be extended until all of them reach the same number based on the *least common multiple number*.

Example 9.3 Let $X = \{x_1, x_2\}$ be the reference set, $h_{\mathbf{A}^{\mathfrak{S}}}(x_1) = (u_{\mathbf{A}^{\mathfrak{S}}}(x_1), v_{\mathbf{A}^{\mathfrak{S}}}(x_1)) = (\langle[s_1, s_4], \{s_2, s_5\}\rangle, \langle[s_1, s_4], \{s_2, s_6\}\rangle)$ and $h_{\mathbf{A}^{\mathfrak{S}}}(x_2) = (u_{\mathbf{A}^{\mathfrak{S}}}(x_2), v_{\mathbf{A}^{\mathfrak{S}}}(x_2)) = (\langle[s_2, s_3], \{s_1, s_3, s_5\}\rangle, \langle[s_1, s_7], \{s_2, s_5, s_6\}\rangle)$ be two LICHVs. Then the unified LICHVs $h_{\mathbf{A}^{\mathfrak{S}}}(x_1)$ and $h_{\mathbf{A}^{\mathfrak{S}}}(x_2)$ can be reconsidered as

$$\begin{aligned}\mathbf{A}^{\mathfrak{S}} = \{&(\langle[s_1, s_4], \{s_2, s_2, s_2, s_5, s_5, s_5\}\rangle, \langle[s_1, s_4], \{s_2, s_2, s_2, s_6, s_6, s_6\}\rangle),\\ &(\langle[s_2, s_3], \{s_1, s_1, s_3, s_3, s_5, s_5\}\rangle, \langle[s_1, s_7], \{s_2, s_2, s_5, s_5, s_6, s_6\}\rangle)\}.\end{aligned}$$

Assume that $\mathfrak{S} = \{s_\alpha \mid \alpha \in [0, \tau]\}$ where any linguistic term set $s_0 \leq s_\alpha \leq s_\tau$ is continuous, and $h_{\mathbf{A}^{\mathfrak{S}}} = (u_{\mathbf{A}^{\mathfrak{S}}}, v_{\mathbf{A}^{\mathfrak{S}}}) = (\langle[s^L_{\gamma_{\mathbf{A}^{\mathfrak{S}}}}, s^U_{\gamma_{\mathbf{A}^{\mathfrak{S}}}}], \bigcup_{s_{\gamma_{\mathbf{A}^{\mathfrak{S}}}} \in u_{\mathbf{A}^{\mathfrak{S}}}} \{s_{\gamma_{\mathbf{A}^{\mathfrak{S}}}}\}\rangle, \langle[s^L_{\eta_{\mathbf{A}^{\mathfrak{S}}}}, s^U_{\eta_{\mathbf{A}^{\mathfrak{S}}}}], \bigcup_{s_{\eta_{\mathbf{A}^{\mathfrak{S}}}} \in v_{\mathbf{A}^{\mathfrak{S}}}} \{s_{\eta_{\mathbf{A}^{\mathfrak{S}}}}\}\rangle)$, $h_{\mathbf{A}_1^{\mathfrak{S}}} = (u_{\mathbf{A}_1^{\mathfrak{S}}}, v_{\mathbf{A}_1^{\mathfrak{S}}}) = (\langle[s^L_{\gamma_{\mathbf{A}_1^{\mathfrak{S}}}}, s^U_{\gamma_{\mathbf{A}_1^{\mathfrak{S}}}}], \bigcup_{s_{\gamma_{\mathbf{A}_1^{\mathfrak{S}}}} \in u_{\mathbf{A}_1^{\mathfrak{S}}}} \{s_{\gamma_{\mathbf{A}_1^{\mathfrak{S}}}}\}\rangle, \langle[s^L_{\eta_{\mathbf{A}_1^{\mathfrak{S}}}}, s^U_{\eta_{\mathbf{A}_1^{\mathfrak{S}}}}], \bigcup_{s_{\eta_{\mathbf{A}_1^{\mathfrak{S}}}} \in v_{\mathbf{A}_1^{\mathfrak{S}}}} \{s_{\eta_{\mathbf{A}_1^{\mathfrak{S}}}}\}\rangle)$ and $h_{\mathbf{A}_2^{\mathfrak{S}}} = (u_{\mathbf{A}_2^{\mathfrak{S}}}, v_{\mathbf{A}_2^{\mathfrak{S}}}) = (\langle[s^L_{\gamma_{\mathbf{A}_2^{\mathfrak{S}}}}, s^U_{\gamma_{\mathbf{A}_2^{\mathfrak{S}}}}], \bigcup_{s_{\gamma_{\mathbf{A}_2^{\mathfrak{S}}}} \in u_{\mathbf{A}_2^{\mathfrak{S}}}} \{s_{\gamma_{\mathbf{A}_2^{\mathfrak{S}}}}\}\rangle, \langle[s^L_{\eta_{\mathbf{A}_2^{\mathfrak{S}}}}, s^U_{\eta_{\mathbf{A}_2^{\mathfrak{S}}}}], \bigcup_{s_{\eta_{\mathbf{A}_2^{\mathfrak{S}}}} \in v_{\mathbf{A}_2^{\mathfrak{S}}}} \{s_{\eta_{\mathbf{A}_2^{\mathfrak{S}}}}\}\rangle)$ are three LICHVs. Then, the following operational laws on LICHVs are defined as [7]:

$$\begin{aligned}h_{\mathbf{A}_1^{\mathfrak{S}}} \oplus h_{\mathbf{A}_2^{\mathfrak{S}}} = \Big(&\Big\langle\big[s^L_{\gamma_{\mathbf{A}_1^{\mathfrak{S}}}}, s^U_{\gamma_{\mathbf{A}_1^{\mathfrak{S}}}}\big] \oplus \big[s^L_{\gamma_{\mathbf{A}_2^{\mathfrak{S}}}}, s^U_{\gamma_{\mathbf{A}_2^{\mathfrak{S}}}}\big],\\ &\bigcup_{s_{\gamma_{\mathbf{A}_1^{\mathfrak{S}}}} \in u_{\mathbf{A}_1^{\mathfrak{S}}}} \{s_{\gamma_{\mathbf{A}_1^{\mathfrak{S}}}}\} \oplus \bigcup_{s_{\gamma_{\mathbf{A}_2^{\mathfrak{S}}}} \in u_{\mathbf{A}_2^{\mathfrak{S}}}} \{s_{\gamma_{\mathbf{A}_2^{\mathfrak{S}}}}\}\Big\rangle,\\ &\Big\langle\big[s^L_{\eta_{\mathbf{A}_1^{\mathfrak{S}}}}, s^U_{\eta_{\mathbf{A}_1^{\mathfrak{S}}}}\big] \otimes \big[s^L_{\eta_{\mathbf{A}_2^{\mathfrak{S}}}}, s^U_{\eta_{\mathbf{A}_2^{\mathfrak{S}}}}\big],\\ &\bigcup_{s_{\eta_{\mathbf{A}_1^{\mathfrak{S}}}} \in v_{\mathbf{A}_1^{\mathfrak{S}}}} \{s_{\eta_{\mathbf{A}_1^{\mathfrak{S}}}}\} \otimes \bigcup_{s_{\eta_{\mathbf{A}_2^{\mathfrak{S}}}} \in v_{\mathbf{A}_2^{\mathfrak{S}}}} \{s_{\eta_{\mathbf{A}_2^{\mathfrak{S}}}}\}\Big\rangle\Big)\\ &\Big(\Big\langle\Big[s^L_{\gamma_{\mathbf{A}_1^{\mathfrak{S}}}+\gamma_{\mathbf{A}_2^{\mathfrak{S}}}-\frac{\gamma_{\mathbf{A}_1^{\mathfrak{S}}}\times\gamma_{\mathbf{A}_2^{\mathfrak{S}}}}{\tau}}, s^U_{\gamma_{\mathbf{A}_1^{\mathfrak{S}}}+\gamma_{\mathbf{A}_2^{\mathfrak{S}}}-\frac{\gamma_{\mathbf{A}_1^{\mathfrak{S}}}\times\gamma_{\mathbf{A}_2^{\mathfrak{S}}}}{\tau}}\Big],\\ &\bigcup_{s_{\gamma_{\mathbf{A}_1^{\mathfrak{S}}}} \in u_{\mathbf{A}_1^{\mathfrak{S}}}, s_{\gamma_{\mathbf{A}_2^{\mathfrak{S}}}} \in u_{\mathbf{A}_2^{\mathfrak{S}}}} \Big\{s_{\gamma_{\mathbf{A}_1^{\mathfrak{S}}}+\gamma_{\mathbf{A}_2^{\mathfrak{S}}}-\frac{\gamma_{\mathbf{A}_1^{\mathfrak{S}}}\times\gamma_{\mathbf{A}_2^{\mathfrak{S}}}}{\tau}}\Big\}\Big\rangle,\\ &\Big\langle\Big[s^L_{\frac{\eta_{\mathbf{A}_1^{\mathfrak{S}}}\times\eta_{\mathbf{A}_2^{\mathfrak{S}}}}{\tau}}, s^U_{\frac{\eta_{\mathbf{A}_1^{\mathfrak{S}}}\times\eta_{\mathbf{A}_2^{\mathfrak{S}}}}{\tau}}\Big],\\ &\bigcup_{s_{\eta_{\mathbf{A}_1^{\mathfrak{S}}}} \in v_{\mathbf{A}_1^{\mathfrak{S}}}, s_{\eta_{\mathbf{A}_2^{\mathfrak{S}}}} \in v_{\mathbf{A}_2^{\mathfrak{S}}}} \Big\{s_{\frac{\eta_{\mathbf{A}_1^{\mathfrak{S}}}\times\eta_{\mathbf{A}_2^{\mathfrak{S}}}}{\tau}}\Big\}\Big\rangle\Big);\end{aligned} \tag{9.7}$$

$$
\begin{aligned}
h_{\mathbf{A}_1\mathfrak{S}} \otimes h_{\mathbf{A}_2\mathfrak{S}} = & \Bigg(\Bigg\langle\left[s^L_{\gamma_{\mathbf{A}_1\mathfrak{S}}}, s^U_{\gamma_{\mathbf{A}_1\mathfrak{S}}}\right] \otimes \left[s^L_{\gamma_{\mathbf{A}_2\mathfrak{S}}}, s^U_{\gamma_{\mathbf{A}_2\mathfrak{S}}}\right], \\
& \bigcup_{s_{\gamma_{\mathbf{A}_1\mathfrak{S}}} \in u_{\mathbf{A}_1\mathfrak{S}}} \{s_{\gamma_{\mathbf{A}_1\mathfrak{S}}}\} \otimes \bigcup_{s_{\gamma_{\mathbf{A}_2\mathfrak{S}}} \in u_{\mathbf{A}_2\mathfrak{S}}} \{s_{\gamma_{\mathbf{A}_2\mathfrak{S}}}\}\Bigg\rangle, \\
& \Bigg\langle\left[s^L_{\eta_{\mathbf{A}_1\mathfrak{S}}}, s^U_{\eta_{\mathbf{A}_1\mathfrak{S}}}\right] \oplus \left[s^L_{\eta_{\mathbf{A}_2\mathfrak{S}}}, s^U_{\eta_{\mathbf{A}_2\mathfrak{S}}}\right], \\
& \bigcup_{s_{\eta_{\mathbf{A}_1\mathfrak{S}}} \in v_{\mathbf{A}_1\mathfrak{S}}} \{s_{\eta_{\mathbf{A}_1\mathfrak{S}}}\} \oplus \bigcup_{s_{\eta_{\mathbf{A}_2\mathfrak{S}}} \in v_{\mathbf{A}_2\mathfrak{S}}} \{s_{\eta_{\mathbf{A}_2\mathfrak{S}}}\}\Bigg\rangle\Bigg) \\
& \Bigg(\Bigg\langle\left[s^L_{\frac{\gamma_{\mathbf{A}_1\mathfrak{S}} \times \gamma_{\mathbf{A}_2\mathfrak{S}}}{\tau}}, s^U_{\frac{\gamma_{\mathbf{A}_1\mathfrak{S}} \times \gamma_{\mathbf{A}_2\mathfrak{S}}}{\tau}}\right], \\
& \bigcup_{s_{\gamma_{\mathbf{A}_1\mathfrak{S}}} \in u_{\mathbf{A}_1\mathfrak{S}}, s_{\gamma_{\mathbf{A}_2\mathfrak{S}}} \in u_{\mathbf{A}_2\mathfrak{S}}} \left\{s_{\frac{\gamma_{\mathbf{A}_1\mathfrak{S}} \times \gamma_{\mathbf{A}_2\mathfrak{S}}}{\tau}}\right\}\Bigg\rangle, \\
& \Bigg\langle\Bigg[s^L_{\eta_{\mathbf{A}_1\mathfrak{S}} + \eta_{\mathbf{A}_2\mathfrak{S}} - \frac{\eta_{\mathbf{A}_1\mathfrak{S}} \times \eta_{\mathbf{A}_2\mathfrak{S}}}{\tau}}, \\
& s^U_{\eta_{\mathbf{A}_1\mathfrak{S}} + \eta_{\mathbf{A}_2\mathfrak{S}} - \frac{\eta_{\mathbf{A}_1\mathfrak{S}} \times \eta_{\mathbf{A}_2\mathfrak{S}}}{\tau}}\Bigg], \\
& \bigcup_{s_{\eta_{\mathbf{A}_1\mathfrak{S}}} \in v_{\mathbf{A}_1\mathfrak{S}}, s_{\eta_{\mathbf{A}_2\mathfrak{S}}} \in v_{\mathbf{A}_2\mathfrak{S}}} \left\{s_{\eta_{\mathbf{A}_1\mathfrak{S}} + \eta_{\mathbf{A}_2\mathfrak{S}} - \frac{\eta_{\mathbf{A}_1\mathfrak{S}} \times \eta_{\mathbf{A}_2\mathfrak{S}}}{\tau}}\right\}\Bigg\rangle\Bigg);
\end{aligned}
\tag{9.8}
$$

$$
\begin{aligned}
\lambda h_{\mathbf{A}\mathfrak{S}} = & \Bigg(\Bigg\langle[s^L_{\tau-\tau(1-\frac{\gamma_{\mathbf{A}\mathfrak{S}}}{\tau})^\lambda}, s^U_{\tau-\tau(1-\frac{\gamma_{\mathbf{A}\mathfrak{S}}}{\tau})^\lambda}], \bigcup_{s_{\gamma_{\mathbf{A}\mathfrak{S}}} \in u_{\mathbf{A}\mathfrak{S}}} \left\{s_{\tau-\tau(1-\frac{\gamma_{\mathbf{A}\mathfrak{S}}}{\tau})^\lambda}\right\}\Bigg\rangle, \\
& \Bigg\langle\left[s^L_{\tau(\frac{\eta_{\mathbf{A}\mathfrak{S}}}{\tau})^\lambda}, s^U_{\tau(\frac{\eta_{\mathbf{A}\mathfrak{S}}}{\tau})^\lambda}\right], \bigcup_{s_{\eta_{\mathbf{A}\mathfrak{S}}} \in v_{\mathbf{A}\mathfrak{S}}} \{s_{\tau(\frac{\eta_{\mathbf{A}\mathfrak{S}}}{\tau})^\lambda}\}\Bigg\rangle\Bigg),
\end{aligned}
\tag{9.9}
$$

$$
\begin{aligned}
(h_{\mathbf{A}\mathfrak{S}})^\lambda = & \Bigg(\Bigg\langle\left[s^L_{\tau(\frac{\gamma_{\mathbf{A}\mathfrak{S}}}{\tau})^\lambda}, s^U_{\tau(\frac{\gamma_{\mathbf{A}\mathfrak{S}}}{\tau})^\lambda}\right], \bigcup_{s_{\gamma_{\mathbf{A}\mathfrak{S}}} \in u_{\mathbf{A}\mathfrak{S}}} \{s_{\tau(\frac{\gamma_{\mathbf{A}\mathfrak{S}}}{\tau})^\lambda}\}\Bigg\rangle, \\
& \Bigg\langle\left[s^L_{\tau-\tau(1-\frac{\eta_{\mathbf{A}\mathfrak{S}}}{\tau})^\lambda}, s^U_{\tau-\tau(1-\frac{\eta_{\mathbf{A}\mathfrak{S}}}{\tau})^\lambda}\right], \\
& \bigcup_{s_{\eta_{\mathbf{A}\mathfrak{S}}} \in v_{\mathbf{A}\mathfrak{S}}} \left\{s_{\tau-\tau(1-\frac{\eta_{\mathbf{A}\mathfrak{S}}}{\tau})^\lambda}\right\}\Bigg\rangle\Bigg).
\end{aligned}
\tag{9.10}
$$

Remark 9.4 There are other extensions of cubic hesitant fuzzy set that can be found in the existing literature, such as, triangular neutrosophic cubic linguistic hesitant fuzzy set in [3].

References

1. A. Fahmi, F. Amin, F. Smarandache, M. Khan, N. Hassan, Triangular cubic hesitant fuzzy Einstein hybrid weighted averaging operator and its application to decision making. Symmetry **10**, 658 (2018)
2. A. Fahmi, M. Aslam, F. Ali, A. Almahdi, F. Amin, New type of cancer patients based on triangular cubic hesitant fuzzy TOPSIS method. Int. J. Biomath. **13** (2020). https://doi.org/10.1142/s1793524520500023
3. A. Fahmi, M. Aslam, M. Riaz, New approach of triangular neutrosophic cubic linguistic hesitant fuzzy aggregation operators. Granular Comput. **5**, 527–543 (2020)
4. J. Fu, J. Ye, W. Cui, An evaluation method of risk grades for prostate cancer using similarity measure of cubic hesitant fuzzy sets. J. Biomed. Inf. **87**, 131–137 (2018)
5. Y.B. Jun, C.S. Kim, K.O. Yang, Cubic sets. Ann. Fuzzy Math. Inf. **4**, 83–98 (2012)
6. T. Mahmood, F. Mehmood, Q. Khan, Cubic hesitant fuzzy sets and their applications criteria decision making. Int. J. Algebra Statist. **5**, 19–51 (2016)
7. M. Qiyas, S. Abdullah, Muneeza, A novel approach of linguistic intuitionistic cubic hesitant variables and their application in decision making. Granular Comput. (2020). https://doi.org/10.1007/s41066-020-00225-3
8. R. Yong, A. Zhu, J. Ye, Multiple attribute decision method using similarity measure of cubic hesitant fuzzy sets. J. Intell. Fuzzy Syst. **37**, 1075–1083 (2019)

Chapter 10
Complex Hesitant Fuzzy Set

Abstract In this chapter, we present the concept of complex hesitant fuzzy set, and then, we deal with complex dual hesitant fuzzy set. The last part of this chapter is devoted to presenting the concept of complex dual type 2 hesitant fuzzy set.

10.1 Complex Hesitant Fuzzy Set

Mahmood [3] explored the approach of complex hesitant fuzzy set, which contains truth grades in the form of subset of unit disc in the complex plane. Then, they described the operational laws of complex hesitant fuzzy sets and verified them by the use of a number of numerical examples. Garg et al. [1] represented a set of generalized distance measures for complex hesitant fuzzy sets, applied them to a decision making technique being established under the complex hesitant fuzzy environment.

Suppose that X represents a reference set, and

$$A = \{\langle x, \gamma_A(x)\rangle \mid x \in X\} \tag{10.1}$$

denotes a complex fuzzy set A on X in which the notation $\gamma_A(x) = \rho_{\gamma_A}(x) \times e^{i2\pi(\theta_{\gamma_A}(x))}$ indicates a polar representation of complex-valued truth grade with the two parameters $\rho_{\gamma_A}(x)$ and $\theta_{\gamma_A}(x) \in [0, 1]$ (see [6]). Furthermore, the set $\gamma_A(x)$ is in correspondence with the element $x \in X$.

B. Farhadinia, *Hesitant Fuzzy Set*, Computational Intelligence Methods and Applications, https://doi.org/10.1007/978-981-16-7301-6_10

These findings provide us with a deeper understanding of complex hesitant fuzzy set introduced below.

Definition 10.1 ([3]) Let X be a reference set. A complex hesitant fuzzy set (CHFS) A on X is defined as

$$A = \{\langle x, h_A(x)\rangle \mid x \in X\} = \left\{\left\langle x, \bigcup_{\gamma_A \in h_A(x)} \{\gamma_A\}\right\rangle \mid x \in X\right\}$$

$$= \left\{\left\langle x, \bigcup_{\gamma_A \in h_A(x)} \left\{\rho_{\gamma_A}(x) \times e^{i2\pi(\theta_{\gamma_A}(x))}\right\}\right\rangle \mid x \in X\right\}, \tag{10.2}$$

where $h_A(x)$ is the set of different finite values in [0, 1] which represent the grade of truth for each element $x \in X$ to A.

For the sake of simplicity, $h_A(x)$ is called the complex hesitant fuzzy number (CHFN).

Remark 10.1 Throughout this book, the set of all CHFSs on the reference set X is denoted by $\mathbb{CHFS}(X)$.

Example 10.1 Let $X = \{x_1, x_2\}$ be the reference set,
$h_A(x_1) = \{0.1e^{i2\pi(0.3)}, 0.8e^{i2\pi(0.2)}, 0.3e^{i2\pi(0.5)}\}$
and
$h_A(x_2) = \{0.2e^{i2\pi(0.4)}, 0.4e^{i2\pi(0.7)}, 0.1e^{i2\pi(0.5)}\}$ be the CHFEs of x_i $(i = 1, 2)$ to a set A, respectively. Then A can be considered as a CHFS, i.e.,

$$A = \left\{\left\langle x_1, \left\{0.1e^{i2\pi(0.3)}, 0.8e^{i2\pi(0.2)}, 0.3e^{i2\pi(0.5)}\right\}\right\rangle, \left\langle x_2, \left\{0.2e^{i2\pi(0.4)}, 0.4e^{i2\pi(0.7)}, 0.1e^{i2\pi(0.5)}\right\}\right\rangle\right\}.$$

Given a fixed $x \in X$, some operational laws on CHFNs $h_A(x) = \bigcup_{\gamma_A \in h_A(x)}\{\gamma_A\} = \bigcup_{\gamma_A \in h_A(x)}\{\rho_{\gamma_A}(x) \times e^{i2\pi(\theta_{\gamma_A}(x))}\}$, $h_{A_1}(x) = \bigcup_{\gamma_{A_1} \in h_{A_1}(x)}\{\gamma_{A_1}\} = \bigcup_{\gamma_{A_1} \in h_{A_1}(x)}\{\rho_{\gamma_{A_1}}(x) \times e^{i2\pi(\theta_{\gamma_{A_1}}(x))}\}$ and $h_{A_2}(x) = \bigcup_{\gamma_{A_2} \in h_{A_2}(x)}\{\gamma_{A_2}\} = \bigcup_{\gamma_{A_2} \in h_{A_2}(x)}\{\rho_{\gamma_{A_2}}(x) \times e^{i2\pi(\theta_{\gamma_{A_2}}(x))}\}$ are defined as follows (see [3]):

$$h_A{}^c = \bigcup_{\gamma_{A^c} \in h_{A^c}(x)} \{\gamma_{A^c}\} = \bigcup_{\gamma_A \in h_A(x)} \left\{(1 - \rho_{\gamma_A}(x)) \times e^{i2\pi(1-\theta_{\gamma_A}(x))}\right\}; \tag{10.3}$$

$$h_{A_1} \cup h_{A_2} = \bigcup_{\gamma_{A_1} \in h_{A_1}(x), \gamma_{A_2} \in h_{A_2}(x)} \left\{\max\{\rho_{\gamma_{A_1}}(x), \rho_{\gamma_{A_2}}(x)\} \times e^{i2\pi(\max\{\theta_{\gamma_{A_1}}(x), \theta_{\gamma_{A_2}}(x)\})}\right\}, \tag{10.4}$$

$$h_{A_1} \cap h_{A_2} = \bigcup_{\gamma_{A_1} \in h_{A_1}(x), \gamma_{A_2} \in h_{A_2}(x)} \left\{ \min\{\rho_{\gamma_{A_1}}(x), \rho_{\gamma_{A_2}}(x)\} \times e^{i2\pi(\min\{\theta_{\gamma_{A_1}}(x), \theta_{\gamma_{A_2}}(x)\})} \right\}. \tag{10.5}$$

10.2 Complex Dual Hesitant Fuzzy Set

Mahmood et al. [4] introduced the concept of complex dual hesitant fuzzy set as an assortment of complex fuzzy set and dual hesitant fuzzy set. Complex dual hesitant fuzzy set contains the grades of membership and non-membership in the form of a complex number which belongs to the complex plane in a unit disc. Mahmood et al. [4] studied a number of complex dual hesitant fuzzy operational laws together with studying two classes of vector similarity measures and hybrid vector similarity measures for complex dual hesitant fuzzy sets. They also described the notion of complex interval-valued dual hesitant fuzzy set.

Definition 10.2 ([4]) Let X be a reference set. A complex dual hesitant fuzzy set (CDHFS) $\widetilde{A}$ on X is defined in terms of $u_{\widetilde{A}}(x)$ and $v_{\widetilde{A}}(x)$ as follows:

$$\widetilde{A} = \{\langle x, u_{\widetilde{A}}(x), v_{\widetilde{A}}(x)\rangle \mid x \in X\}, \tag{10.6}$$

where

$$u_{\widetilde{A}}(x) = \bigcup_{\gamma_{\widetilde{A}} \in u_{\widetilde{A}}(x)} \{\gamma_{\widetilde{A}}\} = \bigcup_{\gamma_{\widetilde{A}} \in u_{\widetilde{A}}(x)} \left\{\rho_{\gamma_{\widetilde{A}}}(x) \times e^{i2\pi(\theta_{\gamma_{\widetilde{A}}}(x))}\right\};$$

$$v_{\widetilde{A}}(x) = \bigcup_{\eta_{\widetilde{A}} \in v_{\widetilde{A}}(x)} \{\eta_{\widetilde{A}}\} = \bigcup_{\eta_{\widetilde{A}} \in v_{\widetilde{A}}(x)} \left\{\rho_{\eta_{\widetilde{A}}}(x) \times e^{i2\pi(\theta_{\eta_{\widetilde{A}}}(x))}\right\}.$$

Here, $u_{\widetilde{A}}(x)$ denotes the complex-valued membership grades, and $v_{\widetilde{A}}(x)$ indicates the complex-valued non-membership grades being subsets of a unit disc in the complex plane with the conditions $\rho^+_{\gamma_{\widetilde{A}}} = \bigcup_{x \in X} \max_{\gamma_{\widetilde{A}} \in u_{\widetilde{A}}(x)}\{\rho_{\gamma_{\widetilde{A}}}(x)\}$, $\theta^+_{\gamma_{\widetilde{A}}} = \bigcup_{x \in X} \max_{\gamma_{\widetilde{A}} \in u_{\widetilde{A}}(x)}\{\theta_{\gamma_{\widetilde{A}}}(x)\}$, $\rho^+_{\eta_{\widetilde{A}}} = \bigcup_{x \in X} \max_{\eta_{\widetilde{A}} \in v_{\widetilde{A}}(x)}\{\rho_{\eta_{\widetilde{A}}}(x)\}$,
and $\theta^+_{\eta_{\widetilde{A}}} = \bigcup_{x \in X} \max_{\eta_{\widetilde{A}} \in v_{\widetilde{A}}(x)}\{\theta_{\eta_{\widetilde{A}}}(x)\}$. Then we will find that

$$0 \le \rho_{\gamma_{\widetilde{A}}}, \theta_{\gamma_{\widetilde{A}}}, \rho_{\eta_{\widetilde{A}}}, \theta_{\eta_{\widetilde{A}}} \le 1, \quad 0 \le \rho^+_{\gamma_{\widetilde{A}}} + \rho^+_{\eta_{\widetilde{A}}} \le 1, \quad 0 \le \theta^+_{\gamma_{\widetilde{A}}} + \theta^+_{\eta_{\widetilde{A}}} \le 1.$$

For the sake of simplicity, $h_{\widetilde{A}}(x) = \langle u_{\widetilde{A}}(x), v_{\widetilde{A}}(x)\rangle$ is called the complex dual hesitant fuzzy number (CDHFN).

Remark 10.2 Throughout this book, the set of all CDHFSs on the reference set X is denoted by $\mathbb{CDHFS}(X)$.

Example 10.2 Let $X = \{x_1, x_2\}$ be the reference set,
$h_{\widetilde{A}}(x_1) = \langle u_{\widetilde{A}}(x_1), v_{\widetilde{A}}(x_1)\rangle = \langle\{0.1e^{i2\pi(0.3)}, 0.8e^{i2\pi(0.2)}, 0.3e^{i2\pi(0.5)}\}, \{0.2e^{i2\pi(0.4)}, 0.3e^{i2\pi(0.3)}\}\rangle$
and
$h_{\widetilde{A}}(x_2) = \langle u_{\widetilde{A}}(x_2), v_{\widetilde{A}}(x_2)\rangle = \langle\{0.2e^{i2\pi(0.4)}, 0.4e^{i2\pi(0.7)}, 0.1e^{i2\pi(0.5)}\}, \{0.6e^{i2\pi(0.2)}, 0.3e^{i2\pi(0.3)}, 0.2e^{i2\pi(0.5)}\}\rangle$
be the CDHFNs of x_i $(i = 1, 2)$ to a set $\widetilde{A}$, respectively. Then $\widetilde{A}$ can be considered as a CDHFS, i.e.,

$$\begin{aligned}\widetilde{A} = \Big\{&\langle x_1, \langle\{0.1e^{i2\pi(0.3)}, 0.8e^{i2\pi(0.2)}, 0.3e^{i2\pi(0.5)}\}, \{0.2e^{i2\pi(0.4)}, 0.3e^{i2\pi(0.3)}\}\rangle,\\ &\langle x_2, \langle\{0.2e^{i2\pi(0.4)}, 0.4e^{i2\pi(0.7)}, 0.1e^{i2\pi(0.5)}\},\\ &\{0.6e^{i2\pi(0.2)}, 0.3e^{i2\pi(0.3)}, 0.2e^{i2\pi(0.5)}\}\rangle\rangle\Big\}.\end{aligned}$$

Given a fixed $x \in X$, some operational laws on CDHFNs $h_{\widetilde{A}} = \langle u_{\widetilde{A}}, v_{\widetilde{A}}\rangle = \langle \bigcup_{\gamma_{\widetilde{A}} \in u_{\widetilde{A}}(x)}\{\gamma_{\widetilde{A}}\}, \bigcup_{\eta_{\widetilde{A}} \in v_{\widetilde{A}}}\{\eta_{\widetilde{A}}\}\rangle = \langle \bigcup_{\gamma_{\widetilde{A}} \in u_{\widetilde{A}}}\{\rho_{\gamma_{\widetilde{A}}} \times e^{i2\pi(\theta_{\gamma_{\widetilde{A}}})}\}, \bigcup_{\eta_{\widetilde{A}} \in v_{\widetilde{A}}}\{\rho_{\eta_{\widetilde{A}}} \times e^{i2\pi(\theta_{\eta_{\widetilde{A}}})}\}\rangle$,
$h_{\widetilde{A}_1} = \langle u_{\widetilde{A}_1}, v_{\widetilde{A}_1}\rangle = \langle \bigcup_{\gamma_{\widetilde{A}_1} \in u_{\widetilde{A}_1}(x)}\{\gamma_{\widetilde{A}_1}\}, \bigcup_{\eta_{\widetilde{A}_1} \in v_{\widetilde{A}_1}}\{\eta_{\widetilde{A}_1}\}\rangle = \langle \bigcup_{\gamma_{\widetilde{A}_1} \in u_{\widetilde{A}_1}}\{\rho_{\gamma_{\widetilde{A}_1}} \times e^{i2\pi(\theta_{\gamma_{\widetilde{A}_1}})}\}, \bigcup_{\eta_{\widetilde{A}_1} \in v_{\widetilde{A}_1}}\{\rho_{\eta_{\widetilde{A}_1}} \times e^{i2\pi(\theta_{\eta_{\widetilde{A}_1}})}\}\rangle$ and
$h_{\widetilde{A}_2} = \langle u_{\widetilde{A}_2}, v_{\widetilde{A}_2}\rangle = \langle \bigcup_{\gamma_{\widetilde{A}_2} \in u_{\widetilde{A}_2}(x)}\{\gamma_{\widetilde{A}_2}\}, \bigcup_{\eta_{\widetilde{A}_2} \in v_{\widetilde{A}_2}}\{\eta_{\widetilde{A}_2}\}\rangle = \langle \bigcup_{\gamma_{\widetilde{A}_2} \in u_{\widetilde{A}_2}}\{\rho_{\gamma_{\widetilde{A}_2}} \times e^{i2\pi(\theta_{\gamma_{\widetilde{A}_2}})}\}, \bigcup_{\eta_{\widetilde{A}_2} \in v_{\widetilde{A}_2}}\{\rho_{\eta_{\widetilde{A}_2}} \times e^{i2\pi(\theta_{\eta_{\widetilde{A}_2}})}\}\rangle$ are defined as the following (see [4]):

$$h^c_{\widetilde{A}} = \left\langle v_{\widetilde{A}}(x), u_{\widetilde{A}}(x)\right\rangle = \left\langle \bigcup_{\eta_{\widetilde{A}} \in v_{\widetilde{A}}} \left\{\rho_{\eta_{\widetilde{A}}} \times e^{i2\pi(\theta_{\eta_{\widetilde{A}}})}\right\}, \bigcup_{\gamma_{\widetilde{A}} \in u_{\widetilde{A}}} \left\{\rho_{\gamma_{\widetilde{A}}} \times e^{i2\pi(\theta_{\gamma_{\widetilde{A}}})}\right\}\right\rangle; \tag{10.7}$$

$$\begin{aligned}h_{\widetilde{A}_1} \cup h_{\widetilde{A}_2} = \Bigg\langle &\bigcup_{\gamma_{\widetilde{A}_1} \in u_{\widetilde{A}_1}; \gamma_{\widetilde{A}_2} \in u_{\widetilde{A}_2}} \left\{\max\left\{\rho_{\gamma_{\widetilde{A}_1}}, \rho_{\gamma_{\widetilde{A}_2}}\right\} \times e^{i2\pi\left(\max\left\{\theta_{\gamma_{\widetilde{A}_1}}, \theta_{\gamma_{\widetilde{A}_2}}\right\}\right)}\right\},\\ &\bigcup_{\eta_{\widetilde{A}_1} \in v_{\widetilde{A}_1}; \eta_{\widetilde{A}_2} \in v_{\widetilde{A}_2}} \left\{\min\{\rho_{\eta_{\widetilde{A}_1}}, \rho_{\eta_{\widetilde{A}_2}}\} \times e^{i2\pi\left(\max\left\{\theta_{\eta_{\widetilde{A}_1}}, \theta_{\eta_{\widetilde{A}_2}}\right\}\right)}\right\}\Bigg\rangle;\end{aligned} \tag{10.8}$$

$$\begin{aligned}h_{\widetilde{A}_1} \cap h_{\widetilde{A}_2} = \Bigg\langle &\bigcup_{\gamma_{\widetilde{A}_1} \in u_{\widetilde{A}_1}; \gamma_{\widetilde{A}_2} \in u_{\widetilde{A}_2}} \left\{\min\{\rho_{\gamma_{\widetilde{A}_1}}, \rho_{\gamma_{\widetilde{A}_2}}\} \times e^{i2\pi\left(\max\left\{\theta_{\gamma_{\widetilde{A}_1}}, \theta_{\gamma_{\widetilde{A}_2}}\right\}\right)}\right\},\\ &\bigcup_{\eta_{\widetilde{A}_1} \in v_{\widetilde{A}_1}; \eta_{\widetilde{A}_2} \in v_{\widetilde{A}_2}} \left\{\max\{\rho_{\eta_{\widetilde{A}_1}}, \rho_{\eta_{\widetilde{A}_2}}\} \times e^{i2\pi\left(\max\left\{\theta_{\eta_{\widetilde{A}_1}}, \theta_{\eta_{\widetilde{A}_2}}\right\}\right)}\right\}\Bigg\rangle.\end{aligned} \tag{10.9}$$

10.3 Complex Dual Type 2 Hesitant Fuzzy Set

Mahmood et al. [5] defined the concept of complex dual type 2 hesitant fuzzy set by composing the grades of truth falsity. The grades of truth and falsity contain the grades of primary and secondary parts in the form of polar coordinates. In continue, Mahmood et al. [5] proposed some complex dual type 2 hesitant fuzzy operational laws, and then, they introduced a correlation coefficient and some entropy measures for complex dual type 2 hesitant fuzzy sets.

Definition 10.3 ([5]) Let X be a reference set. A complex dual type 2 hesitant fuzzy set (CDT2HFS) $\mathbf{A}$ on X is defined in terms of $u_{\mathbf{A}}(x)$ and $v_{\mathbf{A}}(x)$ as follows:

$$\mathbf{A} = \{\langle x, u_{\mathbf{A}}(x), v_{\mathbf{A}}(x)\rangle \mid x \in X\}, \tag{10.10}$$

where the grade of complex-valued supporting $u_{\mathbf{A}}(x)$ and the grade of complex-valued supporting against $v_{\mathbf{A}}(x)$ are, respectively, represented by

$$\begin{aligned} u_{\mathbf{A}}(x) &= \bigcup_{\left(\gamma_{\mathbf{A}}^{I}, \gamma_{\mathbf{A}}^{II}\right) \in u_{\mathbf{A}}(x)} \{(\gamma_{\mathbf{A}}^{I}, \gamma_{\mathbf{A}}^{II})\} \\ &= \bigcup_{\left(\gamma_{\mathbf{A}}^{I}, \gamma_{\mathbf{A}}^{II}\right) \in u_{\mathbf{A}}(x)} \{(\rho_{\gamma_{\mathbf{A}}^{I}}(x) \times e^{i2\pi(\theta_{\gamma_{\mathbf{A}}^{I}}(x))}, \rho_{\gamma_{\mathbf{A}}^{II}}(x) \times e^{i2\pi(\theta_{\gamma_{\mathbf{A}}^{II}}(x))})\}; \\ v_{\mathbf{A}}(x) &= \bigcup_{\left(\eta_{\mathbf{A}}^{I}, \eta_{\mathbf{A}}^{II}\right) \in v_{\mathbf{A}}(x)} \{(\eta_{\mathbf{A}}^{I}, \eta_{\mathbf{A}}^{II})\} \\ &= \bigcup_{\left(\eta_{\mathbf{A}}^{I}, \eta_{\mathbf{A}}^{II}\right) \in v_{\mathbf{A}}(x)} \{(\rho_{\eta_{\mathbf{A}}^{I}}(x) \times e^{i2\pi\left(\theta_{\eta_{\mathbf{A}}^{I}}(x)\right)}, \rho_{\eta_{\mathbf{A}}^{II}}(x) \times e^{i2\pi\left(\theta_{\eta_{\mathbf{A}}^{II}}(x)\right)})\}. \end{aligned}$$

Here, the following terms are defined $\rho^{+}_{\gamma_{\mathbf{A}}^{I}} = \bigcup_{x \in X} \max_{\gamma_{\mathbf{A}}^{I} \in u_{\mathbf{A}}(x)}\{\rho_{\gamma_{\mathbf{A}}^{I}}(x)\}$, $\rho^{+}_{\gamma_{\mathbf{A}}^{II}} = \bigcup_{x \in X} \max_{\gamma_{\mathbf{A}}^{II} \in u_{\mathbf{A}}(x)}\{\rho_{\gamma_{\mathbf{A}}^{II}}(x)\}$, $\theta^{+}_{\gamma_{\mathbf{A}}^{I}} = \bigcup_{x \in X} \max_{\gamma_{\mathbf{A}}^{I} \in u_{\mathbf{A}}(x)}\{\theta_{\gamma_{\mathbf{A}}^{I}}(x)\}$, $\theta^{+}_{\gamma_{\mathbf{A}}^{II}} = \bigcup_{x \in X} \max_{\gamma_{\mathbf{A}}^{II} \in u_{\mathbf{A}}(x)}\{\theta_{\gamma_{\mathbf{A}}^{II}}(x)\}$, $\rho^{+}_{\eta_{\mathbf{A}}^{I}} = \bigcup_{x \in X} \max_{\eta_{\mathbf{A}}^{I} \in v_{\mathbf{A}}(x)}\{\rho_{\eta_{\mathbf{A}}^{I}}(x)\}$, $\rho^{+}_{\eta_{\mathbf{A}}^{II}} = \bigcup_{x \in X} \max_{\eta_{\mathbf{A}}^{II} \in v_{\mathbf{A}}(x)}\{\rho_{\eta_{\mathbf{A}}^{II}}(x)\}$, $\theta^{+}_{\eta_{\mathbf{A}}^{I}} = \bigcup_{x \in X} \max_{\eta_{\mathbf{A}}^{I} \in v_{\mathbf{A}}(x)}\{\theta_{\eta_{\mathbf{A}}^{I}}(x)\}$, and $\theta^{+}_{\eta_{\mathbf{A}}^{II}} = \bigcup_{x \in X} \max_{\eta_{\mathbf{A}}^{II} \in v_{\mathbf{A}}(x)}\{\theta_{\eta_{\mathbf{A}}^{II}}(x)\}$ which satisfy

$$0 \le \rho_{\gamma_{\mathbf{A}}^{I}}, \theta_{\gamma_{\mathbf{A}}^{I}}, \rho_{\eta_{\mathbf{A}}^{I}}, \theta_{\eta_{\mathbf{A}}^{I}} \le 1, \quad 0 \le \rho^{+}_{\gamma_{\mathbf{A}}^{I}} + \rho^{+}_{\eta_{\mathbf{A}}^{I}} \le 1, \quad 0 \le \theta^{+}_{\gamma_{\mathbf{A}}^{I}} + \theta^{+}_{\eta_{\mathbf{A}}^{I}} \le 1;$$

$$0 \le \rho_{\gamma_{\mathbf{A}}^{II}}, \theta_{\gamma_{\mathbf{A}}^{II}}, \rho_{\eta_{\mathbf{A}}^{II}}, \theta_{\eta_{\mathbf{A}}^{II}} \le 1, \quad 0 \le \rho^{+}_{\gamma_{\mathbf{A}}^{II}} + \rho^{+}_{\eta_{\mathbf{A}}^{II}} \le 1, \quad 0 \le \theta^{+}_{\gamma_{\mathbf{A}}^{II}} + \theta^{+}_{\eta_{\mathbf{A}}^{II}} \le 1.$$

For the sake of simplicity, $h_{\mathbf{A}}(x) = \langle u_{\mathbf{A}}(x), v_{\mathbf{A}}(x)\rangle$ is called the complex dual hesitant fuzzy number (CDT2HFN).

Remark 10.3 Throughout this book, the set of all CDT2HFSs on the reference set X is denoted by $\mathbb{CDT2HFS}(X)$.

Example 10.3 Let $X = \{x_1, x_2\}$ be the reference set, $h_{\mathbf{A}}(x_1) = \langle u_{\mathbf{A}}(x_1), v_{\mathbf{A}}(x_1)\rangle = \langle\{(0.1e^{i2\pi(0.3)}, 0.8e^{i2\pi(0.2)}), (0.3e^{i2\pi(0.5)}, 0.2e^{i2\pi(0.4)})\}$, $\{(0.2e^{i2\pi(0.4)}, 0.3e^{i2\pi(0.3)})\}\rangle$ and $h_{\mathbf{A}}(x_2) = \langle u_{\mathbf{A}}(x_2), v_{\mathbf{A}}(x_2)\rangle = \langle\{(0.4e^{i2\pi(0.2)}, 0.3e^{i2\pi(0.1)}), (0.3e^{i2\pi(0.5)}, 0.2e^{i2\pi(0.4)})\}$, $\{(0.5e^{i2\pi(0.5)}, 0.1e^{i2\pi(0.6)}), (0.6e^{i2\pi(0.5)}, 0.7e^{i2\pi(0.2)})\}\rangle$ be the CDT2HFNs of x_i $(i = 1, 2)$ to a set $\mathbf{A}$, respectively. Then $\mathbf{A}$ can be considered as an CDT2HFS, i.e.,

$$\begin{aligned}\mathbf{A} = \Big\{&\langle x_1, \langle\{\big(0.1e^{i2\pi(0.3)}, 0.8e^{i2\pi(0.2)}\big), \big(0.3e^{i2\pi(0.5)}, 0.2e^{i2\pi(0.4)}\big)\},\\ &\{\big(0.2e^{i2\pi(0.4)}, 0.3e^{i2\pi(0.3)}\big)\}\rangle, \langle x_2, \langle\{\big(0.4e^{i2\pi(0.2)}, 0.3e^{i2\pi(0.1)}\big),\\ &\big(0.3e^{i2\pi(0.5)}, 0.2e^{i2\pi(0.4)}\big)\}, \{\big(0.5e^{i2\pi(0.5)}, 0.1e^{i2\pi(0.6)}\big),\\ &\big(0.6e^{i2\pi(0.5)}, 0.7e^{i2\pi(0.2)}\big)\}\rangle\Big\}.\end{aligned}$$

Given a fixed $x \in X$, some operational laws on CDT2HFNs $h_{\mathbf{A}} = \langle u_{\mathbf{A}}, v_{\mathbf{A}}\rangle = \langle \bigcup_{(\gamma^I_{\mathbf{A}}, \gamma^{II}_{\mathbf{A}}) \in u_{\mathbf{A}}(x)}\{(\gamma^I_{\mathbf{A}}, \gamma^{II}_{\mathbf{A}})\}, \bigcup_{(\eta^I_{\mathbf{A}}, \eta^{II}_{\mathbf{A}}) \in v_{\mathbf{A}}(x)}\{(\eta^I_{\mathbf{A}}, \eta^{II}_{\mathbf{A}})\}\rangle$ $= \langle \bigcup_{(\gamma^I_{\mathbf{A}}, \gamma^{II}_{\mathbf{A}}) \in u_{\mathbf{A}}(x)}\{(\rho_{\gamma^I_{\mathbf{A}}}(x) \times e^{i2\pi(\theta_{\gamma^I_{\mathbf{A}}}(x))}, \rho_{\gamma^{II}_{\mathbf{A}}}(x) \times e^{i2\pi(\theta_{\gamma^{II}_{\mathbf{A}}}(x))})\}$ $, \bigcup_{(\eta^I_{\mathbf{A}}, \eta^{II}_{\mathbf{A}}) \in v_{\mathbf{A}}(x)}\{(\rho_{\eta^I_{\mathbf{A}}}(x) \times e^{i2\pi(\theta_{\eta^I_{\mathbf{A}}}(x))}, \rho_{\eta^{II}_{\mathbf{A}}}(x) \times e^{i2\pi(\theta_{\eta^{II}_{\mathbf{A}}}(x))})\}\rangle$, $h_{\mathbf{A}_1} = \langle u_{\mathbf{A}_1}, v_{\mathbf{A}_1}\rangle = \langle \bigcup_{(\gamma^I_{\mathbf{A}_1}, \gamma^{II}_{\mathbf{A}_1}) \in u_{\mathbf{A}_1}(x)}\{(\gamma^I_{\mathbf{A}_1}, \gamma^{II}_{\mathbf{A}_1})\}, \bigcup_{(\eta^I_{\mathbf{A}_1}, \eta^{II}_{\mathbf{A}_1}) \in v_{\mathbf{A}_1}(x)}\{(\eta^I_{\mathbf{A}_1}, \eta^{II}_{\mathbf{A}_1})\}\rangle$ $= \langle \bigcup_{(\gamma^I_{\mathbf{A}_1}, \gamma^{II}_{\mathbf{A}_1}) \in u_{\mathbf{A}_1}(x)}\{(\rho_{\gamma^I_{\mathbf{A}_1}}(x) \times e^{i2\pi(\theta_{\gamma^I_{\mathbf{A}_1}}(x))}, \rho_{\gamma^{II}_{\mathbf{A}_1}}(x) \times e^{i2\pi(\theta_{\gamma^{II}_{\mathbf{A}_1}}(x))})\}$ $, \bigcup_{(\eta^I_{\mathbf{A}_1}, \eta^{II}_{\mathbf{A}_1}) \in v_{\mathbf{A}_1}(x)}\{(\rho_{\eta^I_{\mathbf{A}_1}}(x) \times e^{i2\pi(\theta_{\eta^I_{\mathbf{A}_1}}(x))}, \rho_{\eta^{II}_{\mathbf{A}_1}}(x) \times e^{i2\pi(\theta_{\eta^{II}_{\mathbf{A}_1}}(x))})\}\rangle$ and $h_{\mathbf{A}_2} = \langle u_{\mathbf{A}_2}, v_{\mathbf{A}_2}\rangle = \langle \bigcup_{(\gamma^I_{\mathbf{A}_2}, \gamma^{II}_{\mathbf{A}_2}) \in u_{\mathbf{A}_2}(x)}\{(\gamma^I_{\mathbf{A}_2}, \gamma^{II}_{\mathbf{A}_2})\}, \bigcup_{(\eta^I_{\mathbf{A}_2}, \eta^{II}_{\mathbf{A}_2}) \in v_{\mathbf{A}_2}(x)}\{(\eta^I_{\mathbf{A}_2}, \eta^{II}_{\mathbf{A}_2})\}\rangle$ $= \langle \bigcup_{(\gamma^I_{\mathbf{A}_2}, \gamma^{II}_{\mathbf{A}_2}) \in u_{\mathbf{A}_2}(x)}\{(\rho_{\gamma^I_{\mathbf{A}_2}}(x) \times e^{i2\pi(\theta_{\gamma^I_{\mathbf{A}_2}}(x))}, \rho_{\gamma^{II}_{\mathbf{A}_2}}(x) \times e^{i2\pi(\theta_{\gamma^{II}_{\mathbf{A}_2}}(x))})\}$ $, \bigcup_{(\eta^I_{\mathbf{A}_2}, \eta^{II}_{\mathbf{A}_2}) \in v_{\mathbf{A}_2}(x)}\{(\rho_{\eta^I_{\mathbf{A}_2}}(x) \times e^{i2\pi(\theta_{\eta^I_{\mathbf{A}_2}}(x))}, \rho_{\eta^{II}_{\mathbf{A}_2}}(x) \times e^{i2\pi(\theta_{\eta^{II}_{\mathbf{A}_2}}(x))})\}\rangle$ are defined as the following [5]:

$$h_{\mathbf{A}_1} \cup h_{\mathbf{A}_2} =$$

$$\Big\langle \bigcup_{(\gamma^I_{\mathbf{A}_i}, \gamma^{II}_{\mathbf{A}_i}) \in u_{\mathbf{A}_i},\ i=1,2} \Big\{\Big(\max\{\rho_{\gamma^I_{\mathbf{A}_1}}, \rho_{\gamma^I_{\mathbf{A}_2}}\} \times e^{i2\pi(\max\{\theta_{\gamma^I_{\mathbf{A}_1}}, \theta_{\gamma^I_{\mathbf{A}_2}}\})},$$

$$\max\{\rho_{\gamma^{II}_{\mathbf{A}_1}}, \rho_{\gamma^{II}_{\mathbf{A}_2}}\} \times e^{i2\pi\left(\max\{\theta_{\gamma^{II}_{\mathbf{A}_1}}, \theta_{\gamma^{II}_{\mathbf{A}_2}}\}\right)}\Big)\Big\},$$

$$\bigcup_{(\eta^{I}_{\mathbf{A}_i}, \eta^{II}_{\mathbf{A}_i}) \in v_{\mathbf{A}_i},\ i=1,2} \Big\{\Big(\min\{\rho_{\eta^{I}_{\mathbf{A}_1}}, \rho_{\eta^{I}_{\mathbf{A}_2}}\} \times e^{i2\pi(\min\{\theta_{\eta^{I}_{\mathbf{A}_1}}, \theta_{\eta^{I}_{\mathbf{A}_2}}\})},$$

$$\min\{\rho_{\eta^{II}_{\mathbf{A}_1}}, \rho_{\eta^{II}_{\mathbf{A}_2}}\} \times e^{i2\pi(\min\{\theta_{\eta^{II}_{\mathbf{A}_1}}, \theta_{\eta^{II}_{\mathbf{A}_2}}\})}\Big)\Big\}\Big\rangle; \tag{10.11}$$

$$h_{\mathbf{A}_1} \cap h_{\mathbf{A}_2} =$$

$$\Big\langle \bigcup_{(\gamma^{I}_{\mathbf{A}_i}, \gamma^{II}_{\mathbf{A}_i}) \in u_{\mathbf{A}_i},\ i=1,2} \Big\{\Big(\min\{\rho_{\gamma^{I}_{\mathbf{A}_1}}, \rho_{\gamma^{I}_{\mathbf{A}_2}}\} \times e^{i2\pi(\min\{\theta_{\gamma^{I}_{\mathbf{A}_1}}, \theta_{\gamma^{I}_{\mathbf{A}_2}}\})},$$

$$\min\{\rho_{\gamma^{II}_{\mathbf{A}_1}}, \rho_{\gamma^{II}_{\mathbf{A}_2}}\} \times e^{i2\pi(\min\{\theta_{\gamma^{II}_{\mathbf{A}_1}}, \theta_{\gamma^{II}_{\mathbf{A}_2}}\})}\Big)\Big\},$$

$$\bigcup_{(\eta^{I}_{\mathbf{A}_i}, \eta^{II}_{\mathbf{A}_i}) \in v_{\mathbf{A}_i},\ i=1,2} \Big\{\Big(\max\{\rho_{\eta^{I}_{\mathbf{A}_1}}, \rho_{\eta^{I}_{\mathbf{A}_2}}\} \times e^{i2\pi(\max\{\theta_{\eta^{I}_{\mathbf{A}_1}}, \theta_{\eta^{I}_{\mathbf{A}_2}}\})},$$

$$\max\{\rho_{\eta^{II}_{\mathbf{A}_1}}, \rho_{\eta^{II}_{\mathbf{A}_2}}\} \times e^{i2\pi\left(\max\{\theta_{\eta^{II}_{\mathbf{A}_1}}, \theta_{\eta^{II}_{\mathbf{A}_2}}\}\right)}\Big)\Big\}\Big\rangle. \tag{10.12}$$

Remark 10.4 The other extensions of complex hesitant fuzzy set can be found in the literature, such as complex q-rung orthopair uncertain linguistic set [7], interval-valued complex single-valued neutrosophic hesitant fuzzy set [2], etc.

References

1. H. Garg, T. Mahmood, U. Rehman, Z. Ali, CHFS: complex hesitant fuzzy sets-their applications to decision making with different and innovative distance measures. CAAI Trans. Intell. Technol. **6**, 93–122 (2021)
2. D. Li, T. Mahmood, Z. Ali, Y. Dong, Decision making based on interval-valued complex single-valued neutrosophic hesitant fuzzy generalized hybrid weighted averaging operators. J. Intell. Fuzzy Syst. **38**, 4359–4401 (2020)
3. T. Mahmood, U. Rehman, Z. Ali, Exponential and non-exponential based generalized similarity measures for complex hesitant fuzzy sets with applications. Fuzzy Inf. Eng. (2020) https://doi.org/10.1080/16168658.2020.1779013
4. T. Mahmood, U. Rehman, Z. Ali, R. Chinram, Jaccard and dice similarity measures based on novel complex dual hesitant fuzzy sets and their applications. Math. Probl. Eng. Article ID 5920432. **2020** (2020)

5. T. Mahmood, Z. Ali, H. Garg, L. Zedam, R. Chinram, Correlation coefficient and entropy measures based on complex dual type-2 hesitant fuzzy sets and their applications. J. Math. Article ID 2568391. **2021** (2021)
6. D. Ramot, R. Milo, M. Friedman, Complex fuzzy sets. IEEE Trans. Fuzzy Syst. **10**, 171–186 (2002)
7. M. Yang, Z. Ali, T. Mahmood, Complex q-rung orthopair uncertain linguistic partitioned Bonferroni mean operators with application in antivirus mask selection. Symmetry **13**, 249 (2021)

Chapter 11
Picture Hesitant Fuzzy Set

Abstract In this chapter, we present the concept of picture hesitant fuzzy set, and then, we deal with interval-valued picture hesitant fuzzy set. The last part of this chapter is devoted to presenting the concept of dual picture hesitant fuzzy set.

11.1 Picture Hesitant Fuzzy Set

Wang and Li [6] firstly represented the concept of picture hesitant fuzzy set by combining the concepts of picture fuzzy set and hesitant fuzzy set. They put forward a set of aggregation operators to deal with multiple criteria decision making whose criteria are related at different priorities. Ullah et al. [5] introduced the concept of GRA technique by the help of picture hesitant fuzzy sets with incomplete weight information. Jan et al. [3] developed a number of picture hesitant fuzzy-based distance measures and made a comparison analysis of the proposed distance measures with other existing ones. Mahmood and Ali [4] defined a fuzzy cross-entropy for picture hesitant fuzzy sets, and applied it to multiple criteria decision making. Ambrin et al. [1] proposed a multiple criteria decision making procedure for supplier selection using a TOPSIS technique under picture hesitant fuzzy environment.

By taking X as a reference set into account, a picture fuzzy set A on X is characterized by three functions $u_A(x)$, $w_A(x)$, and $v_A(x)$ as follows (see [2]):

$$A = \{\langle x, u_A(x), w_A(x), v_A(x)\rangle \mid x \in X\}, \tag{11.1}$$

where $u_A(x)$, $w_A(x)$, and $v_A(x)$ are the sets of some different values in [0, 1] and represent the positive, neutral, and negative membership degrees of the element $x \in X$ to A, respectively. The positive, neutral, and negative membership degrees satisfy

$$0 \le u_A(x) + w_A(x) + v_A(x) \le 1$$

B. Farhadinia, *Hesitant Fuzzy Set*, Computational Intelligence Methods and Applications, https://doi.org/10.1007/978-981-16-7301-6_11

and make a presentation of indeterminacy degree as $\pi_A(x) = 1 - u_A(x) - w_A(x) - v_A(x)$ for any $x \in X$.

Given a fixed $x \in X$, some operational laws on PFSs are defined as follows (see [7]):

$$\begin{aligned}
A^c &= \{\langle x, v_A(x), w_A(x), u_A(x)\rangle \mid x \in X\};\\
A_1 \cup A_2 &= \{\langle x, \max\{u_{A_1}(x), u_{A_2}(x)\}, \min\{w_{A_1}(x), w_{A_2}(x)\},\\
&\qquad \min\{v_{A_1}(x), v_{A_2}(x)\}\rangle \mid x \in X\};\\
A_1 \cap A_2 &= \{\langle x, \min\{u_{A_1}(x), u_{A_2}(x)\}, \max\{w_{A_1}(x), w_{A_2}(x)\},\\
&\qquad \max\{v_{A_1}(x), v_{A_2}(x)\}\rangle \mid x \in X\};\\
A_1 \oplus A_2 &= \{\langle x, u_{A_1}(x) + u_{A_2}(x) - u_{A_1}(x) \times u_{A_2}(x), w_{A_1}(x) \times w_{A_2}(x),\\
&\qquad v_{A_1}(x) \times v_{A_2}(x)\}\rangle \mid x \in X\};\\
A_1 \otimes A_2 &= \{\langle x, u_{A_1}(x) \times u_{A_2}(x), w_{A_1}(x) + w_{A_2}(x) - w_{A_1}(x) \times w_{A_2}(x),\\
&\qquad v_{A_1}(x) + v_{A_2}(x) - v_{A_1}(x) \times v_{A_2}(x)\}\rangle \mid x \in X\};\\
\lambda A &= \{\langle x, 1-(1-u_A(x))^\lambda, (w_A(x))^\lambda, (v_A(x))^\lambda\rangle \mid x \in X\},\ \lambda > 0;\\
A^\lambda &= \{\langle x, (u_A(x))^\lambda, 1-(1-w_A(x))^\lambda, 1-(1-v_A(x))^\lambda\rangle \mid x \in X\},\ \lambda > 0.
\end{aligned}$$

Example 11.1 Let $X = \{x_1, x_2\}$ be the reference set, $h_A(x_1) = \{(0.2, 0.2, 0.3)\}$ and $h_A(x_2) = \{(0.4, 0.1, 0.3)\}$ be the PFEs of x_i $(i = 1, 2)$ to a set A, respectively. Then A can be considered as a PFS, i.e.,

$$A = \{\langle x_1, (0.2, 0.2, 0.3)\rangle, \langle x_2, (0.4, 0.1, 0.3)\rangle\}.$$

Definition 11.1 ([6]) Let X be a reference set. A picture hesitant fuzzy set (PHFS) A on X is defined in terms of three functions $u_A(x)$, $w_A(x)$, and $v_A(x)$ as follows:

$$A = \{\langle x, u_A(x), w_A(x), v_A(x)\rangle \mid x \in X\}, \tag{11.2}$$

where $u_A(x)$, $w_A(x)$, and $v_A(x)$ are the sets of some different values in [0, 1] and represent the possible positive, neutral, and negative membership degrees of the element $x \in X$ to A, respectively.

Here, for all $x \in X$, if we consider $u_A(x) = \bigcup_{\gamma_A \in u_A(x)}\{\gamma_A\}$, $w_A(x) = \bigcup_{\delta_A \in w_A(x)}\{\delta_A\}$, $v_A(x) = \bigcup_{\eta_A \in v_A(x)}\{\eta_A\}$, $\gamma_A^+ \in u_A^+ = \bigcup_{x \in X} \max_{\gamma_A \in u_A(x)}\{\gamma_A\}$, $\delta_A^+ \in w_A^+ = \bigcup_{x \in X} \max_{\delta_A \in w_A(x)}\{\delta_A\}$ and $\eta_A^+ \in v_A^+ = \bigcup_{x \in X} \max_{\eta_A \in v_A(x)}\{\eta_A\}$, then we conclude that

$$0 \le \gamma_A, \delta_A, \eta_A \le 1, \quad 0 \le \gamma_A^+ + \delta_A^+ + \eta_A^+ \le 1.$$

For the sake of simplicity, the pair $h_A(x) = (u_A(x), w_A(x), v_A(x))$ is called the picture hesitant fuzzy element (PHFE).

Remark 11.1 Throughout this book, the set of all PHFSs on the reference set X is denoted by $\mathbb{PHFS}(X)$.

Example 11.2 Let $X = \{x_1, x_2\}$ be the reference set, $h_A(x_1) = (u_A(x_1), w_A(x_1), v_A(x_1)) = (\{0.2, 0.3\}, \{0.1, 0.4\}, \{0.3\})$ and $h_A(x_2) = (u_A(x_2), w_A(x_2), v_A(x_2)) = (\{0.3, 0.4\}, \{0.2\}, \{0.1, 0.2\})$. Then, $h_A(x_i)$ for $i = 1, 2$ are the PHFEs of x_i $(i = 1, 2)$ in the set A, because the relations

$$\gamma_A^+(x_1) = 0.3,\ \delta_A^+(x_1) = 0.4,\ \eta_A^+(x_1) = 0.3;$$
$$\gamma_A^+(x_2) = 0.4,\ \delta_A^+(x_2) = 0.2,\ \eta_A^+(x_2) = 0.2$$

hold true. Thus, A is a PHFS, and it is denoted by

$$A = \{\langle x_1, (\{0.2, 0.3\}, \{0.1, 0.4\}, \{0.3\})\rangle,\ \langle x_2, (\{0.3, 0.4\}, \{0.2\}, \{0.1, 0.2\})\rangle\}.$$

For the PHFEs $h_A = (u_A, w_A, v_A)$, $h_{A_1} = (u_{A_1}, w_{A_1}, v_{A_1})$ and $h_{A_2} = (u_{A_2}, w_{A_2}, v_{A_2})$ the following operations are defined:

$$(h_A)^c = (v_A, w_A, u_A); \tag{11.3}$$
$$h_{A_1} \cup h_{A_2} = (u_{A_1} \cup u_{A_2}, w_{A_1} \cap w_{A_2}, v_{A_1} \cap v_{A_2}); \tag{11.4}$$
$$h_{A_1} \cap h_{A_2} = (u_{A_1} \cap u_{A_2}, w_{A_1} \cup w_{A_2}, v_{A_1} \cup v_{A_2}); \tag{11.5}$$
$$h_{A_1} \oplus h_{A_2} = (u_{A_1} \oplus u_{A_2}, w_{A_1} \otimes w_{A_2}, v_{A_1} \otimes v_{A_2}); \tag{11.6}$$
$$h_{A_1} \otimes h_{A_2} = (u_{A_1} \otimes u_{A_2}, w_{A_1} \oplus w_{A_2}, v_{A_1} \oplus v_{A_2}), \tag{11.7}$$

where

$$0 \le \gamma_A^+ + \delta_A^+ + \eta_A^+ \le 1,\quad 0 \le \gamma_{A_1}^+ + \delta_{A_1}^+ + \eta_{A_1}^+ \le 1,\quad 0 \le \gamma_{A_2}^+ + \delta_{A_2}^+ + \eta_{A_2}^+ \le 1.$$

On the basis of the above operations on PHFEs, some relationships can be further established for such operations on PHFSs as follows (see [6]):

$$A_1 \cup A_2 = \bigcup_{h_{A_1} \in A_1, h_{A_2} \in A_2} h_{A_1} \cup h_{A_2}; \tag{11.8}$$

$$A_1 \cap A_2 = \bigcup_{h_{A_1} \in A_1, h_{A_2} \in A_2} h_{A_1} \cap h_{A_2}; \tag{11.9}$$

$$A_1 \oplus A_2 = \bigcup_{h_{A_1} \in A_1, h_{A_2} \in A_2} h_{A_1} \oplus h_{A_2}; \tag{11.10}$$

$$A_1 \otimes A_2 = \bigcup_{h_{A_1} \in A_1, h_{A_2} \in A_2} h_{A_1} \otimes h_{A_2}. \tag{11.11}$$

11.2 Interval-Valued Picture Hesitant Fuzzy Set

Although there does not exist any contribution so far which deals with interval-valued picture hesitant fuzzy set, it can be defined easily by combing the notions of picture fuzzy set and interval-valued hesitant fuzzy set.

Definition 11.2 Let X be a reference set. An interval-valued picture hesitant fuzzy set (IVPHFS) $\widetilde{A}$ on X is defined in terms of three functions $u_{\widetilde{A}}(x)$, $w_{\widetilde{A}}(x)$, and $v_{\widetilde{A}}(x)$ as follows:

$$\widetilde{A} = \{\langle x, u_{\widetilde{A}}(x), w_{\widetilde{A}}(x), v_{\widetilde{A}}(x)\rangle \mid x \in X\}, \tag{11.12}$$

where $u_{\widetilde{A}}(x)$, $w_{\widetilde{A}}(x)$ and $v_{\widetilde{A}}(x)$ are some different interval values in $[0, 1]$ and represent the possible interval positive, interval neutral, and interval negative membership degrees of the element $x \in X$ to the set $\widetilde{A}$, respectively.

Here, for all $x \in X$, if we take $u_{\widetilde{A}}(x) = \bigcup_{[\gamma_{\widetilde{A}}^L, \gamma_{\widetilde{A}}^U] \in u_{\widetilde{A}}(x)} \{[\gamma_{\widetilde{A}}^L, \gamma_{\widetilde{A}}^U]\}$, $w_{\widetilde{A}}(x) = \bigcup_{[\delta_{\widetilde{A}}^L, \delta_{\widetilde{A}}^U] \in w_{\widetilde{A}}(x)} \{[\delta_{\widetilde{A}}^L, \delta_{\widetilde{A}}^U]\}$, $v_{\widetilde{A}}(x) = \bigcup_{[\eta_{\widetilde{A}}^L, \eta_{\widetilde{A}}^U] \in v_{\widetilde{A}}(x)} \{[\eta_{\widetilde{A}}^L, \eta_{\widetilde{A}}^U]\}$, $\gamma_{\widetilde{A}}^{U+} \in \bigcup_{x \in X} \max\{\gamma_{\widetilde{A}}^U(x)\}$, $\delta_{\widetilde{A}}^{U+} \in \bigcup_{x \in X} \max\{\delta_{\widetilde{A}}^U(x)\}$ and $\eta_{\widetilde{A}}^{U+} \in \bigcup_{x \in X} \max\{\eta_{\widetilde{A}}^U(x)\}$, then we find that

$$0 \le \gamma_{\widetilde{A}}^L, \gamma_{\widetilde{A}}^U, \delta_{\widetilde{A}}^L, \delta_{\widetilde{A}}^U, \eta_{\widetilde{A}}^L, \eta_{\widetilde{A}}^U \le 1, \quad 0 \le \gamma_{\widetilde{A}}^{U+} + \delta_{\widetilde{A}}^{U+} + \eta_{\widetilde{A}}^{U+} \le 1.$$

For the sake of simplicity, $h_{\widetilde{A}}(x) = (u_{\widetilde{A}}(x), w_{\widetilde{A}}(x), v_{\widetilde{A}}(x))$ is called the interval-valued picture hesitant fuzzy element (IVPHFE).

Remark 11.2 Throughout this book, the set of all IVPHFSs on the reference set X is denoted by $\mathbb{IVPHFS}(X)$.

Example 11.3 Let $X = \{x_1, x_2\}$ be the reference set,
$h_{\widetilde{A}}(x_1) = (\bigcup_{[\gamma_{\widetilde{A}}^L, \gamma_{\widetilde{A}}^U] \in u_{\widetilde{A}}(x_1)} \{[\gamma_{\widetilde{A}}^L, \gamma_{\widetilde{A}}^U]\}, \bigcup_{[\delta_{\widetilde{A}}^L, \delta_{\widetilde{A}}^U] \in w_{\widetilde{A}}(x_1)} \{[\delta_{\widetilde{A}}^L, \delta_{\widetilde{A}}^U]\}, \bigcup_{[\eta_{\widetilde{A}}^L, \eta_{\widetilde{A}}^U] \in v_{\widetilde{A}}(x_1)} \{[\eta_{\widetilde{A}}^L, \eta_{\widetilde{A}}^U]\}) = (\{[0.2, 0.4], [0.3, 0.3]\}, \{[0.1, 0.2], [0.2, 0.3]\}, \{[0.2, 0.3]\})$ and
$h_{\widetilde{A}}(x_2) = (\bigcup_{[\gamma_{\widetilde{A}}^L, \gamma_{\widetilde{A}}^U] \in u_{\widetilde{A}}(x_2)} \{[\gamma_{\widetilde{A}}^L, \gamma_{\widetilde{A}}^U]\}, \bigcup_{[\delta_{\widetilde{A}}^L, \delta_{\widetilde{A}}^U] \in w_{\widetilde{A}}(x_1)} \{[\delta_{\widetilde{A}}^L, \delta_{\widetilde{A}}^U]\}, \bigcup_{[\eta_{\widetilde{A}}^L, \eta_{\widetilde{A}}^U] \in v_{\widetilde{A}}(x_2)} \{[\eta_{\widetilde{A}}^L, \eta_{\widetilde{A}}^U]\}) = (\{[0.3, 0.4], [0.1, 0.2]\}, \{[0.3, 0.3], [0.1, 0.3]\}, \{[0.2, 0.2]\})$ be the IVPHFEs of x_i $(i = 1, 2)$ in the set $\widetilde{A}$, respectively. In this case,

$$\gamma^{U+}(x_1) = 0.4, \ \delta^{U+}(x_1) = 0.3, \ \eta^{U+}(x_1) = 0.3;$$
$$\gamma^{U+}(x_2) = 0.4, \ \delta^{U+}(x_2) = 0.3, \ \eta^{U+}(x_2) = 0.2$$

result in

$$\gamma^{U+}(x_1) + \delta^{U+}(x_1) + \eta^{U+}(x_1) = 0.4 + 0.3 + 0.3 \leq 1;$$
$$\gamma^{U+}(x_2) + \delta^{U+}(x_2) + \eta^{U+}(x_2) = 0.4 + 0.3 + 0.2 \leq 1.$$

Thus, $h_{\widetilde{A}}(x_1)$ and $h_{\widetilde{A}}(x_2)$ are two IVPHFEs, and then $\widetilde{A}$ can be considered as a IVPHFS, i.e.,

$$\widetilde{A} = \{\langle x_1, (\{[0.2, 0.4], [0.3, 0.3]\}, \{[0.1, 0.2], [0.2, 0.3]\}, \{[0.2, 0.3]\})\rangle,$$
$$\langle x_2, (\{[0.3, 0.4], [0.1, 0.2]\}, \{[0.3, 0.3], [0.1, 0.3]\}, \{[0.2, 0.2]\})\rangle\}.$$

For the IVPHFEs
$h_{\widetilde{A}} = (u_{\widetilde{A}}, w_{\widetilde{A}}, v_{\widetilde{A}})$
$= (\bigcup_{[\gamma^L_{\widetilde{A}}, \gamma^U_{\widetilde{A}}] \in u_{\widetilde{A}}} \{[\gamma^L_{\widetilde{A}}, \gamma^U_{\widetilde{A}}]\}, \bigcup_{[\delta^L_{\widetilde{A}}, \delta^U_{\widetilde{A}}] \in w_{\widetilde{A}}} \{[\delta^L_{\widetilde{A}}, \delta^U_{\widetilde{A}}]\}, \bigcup_{[\eta^L_{\widetilde{A}}, \eta^U_{\widetilde{A}}] \in v_{\widetilde{A}}} \{[\eta^L_{\widetilde{A}}, \eta^U_{\widetilde{A}}]\}),$
$h_{\widetilde{A}_1} = (u_{\widetilde{A}_1}, w_{\widetilde{A}_1}, v_{\widetilde{A}_1})$
$= (\bigcup_{[\gamma^L_{\widetilde{A}_1}, \gamma^U_{\widetilde{A}_1}] \in u_{\widetilde{A}_1}} \{[\gamma^L_{\widetilde{A}_1}, \gamma^U_{\widetilde{A}_1}]\}, \bigcup_{[\delta^L_{\widetilde{A}_1}, \delta^U_{\widetilde{A}_1}] \in w_{\widetilde{A}_1}} \{[\delta^L_{\widetilde{A}_1}, \delta^U_{\widetilde{A}_1}]\}, \bigcup_{[\eta^L_{\widetilde{A}_1}, \eta^U_{\widetilde{A}_1}] \in v_{\widetilde{A}_1}}$
$\{[\eta^L_{\widetilde{A}_1}, \eta^U_{\widetilde{A}_1}]\})$ and
$h_{\widetilde{A}_2} = (u_{\widetilde{A}_2}, w_{\widetilde{A}_2}, v_{\widetilde{A}_2})$
$= (\bigcup_{[\gamma^L_{\widetilde{A}_2}, \gamma^U_{\widetilde{A}_2}] \in u_{\widetilde{A}_2}} \{[\gamma^L_{\widetilde{A}_2}, \gamma^U_{\widetilde{A}_2}]\}, \bigcup_{[\delta^L_{\widetilde{A}_2}, \delta^U_{\widetilde{A}_2}] \in w_{\widetilde{A}_2}} \{[\delta^L_{\widetilde{A}_2}, \delta^U_{\widetilde{A}_2}]\}, \bigcup_{[\eta^L_{\widetilde{A}_2}, \eta^U_{\widetilde{A}_2}] \in v_{\widetilde{A}_2}}$
$\{[\eta^L_{\widetilde{A}_2}, \eta^U_{\widetilde{A}_2}]\})$ the following operations are defined:

$$\begin{aligned} & h_{\widetilde{A}_1} \oplus h_{\widetilde{A}_2} \\ & = (\bigcup_{[\gamma^L_{\widetilde{A}_1}, \gamma^U_{\widetilde{A}_1}] \in u_{\widetilde{A}_1}, [\gamma^L_{\widetilde{A}_2}, \gamma^U_{\widetilde{A}_2}] \in u_{\widetilde{A}_2}} \\ & \quad \times \{[\gamma^L_{\widetilde{A}_1} + \gamma^L_{\widetilde{A}_2} - \gamma^L_{\widetilde{A}_1} \times \gamma^L_{\widetilde{A}_2}, \gamma^U_{\widetilde{A}_1} + \gamma^U_{\widetilde{A}_2} - \gamma^U_{\widetilde{A}_1} \times \gamma^U_{\widetilde{A}_2}]\}, \\ & \quad \bigcup_{[\delta^L_{\widetilde{A}_1}, \delta^U_{\widetilde{A}_1}] \in w_{\widetilde{A}_1}, [\delta^L_{\widetilde{A}_2}, \delta^U_{\widetilde{A}_2}] \in w_{\widetilde{A}_2}} \{[\delta^L_{\widetilde{A}_1} \times \delta^L_{\widetilde{A}_2}, \delta^U_{\widetilde{A}_1} \times \delta^U_{\widetilde{A}_2}]\}, \\ & \quad \bigcup_{[\eta^L_{\widetilde{A}_1}, \eta^U_{\widetilde{A}_1}] \in v_{\widetilde{A}_1}, [\eta^L_{\widetilde{A}_2}, \eta^U_{\widetilde{A}_2}] \in v_{\widetilde{A}_2}} \{[\eta^L_{\widetilde{A}_1} \times \eta^L_{\widetilde{A}_2}, \eta^U_{\widetilde{A}_1} \times \eta^U_{\widetilde{A}_2}]\}); \end{aligned} \tag{11.13}$$

$$\begin{aligned} & h_{\widetilde{A}_1} \otimes h_{\widetilde{A}_2} \\ & = (\bigcup_{[\gamma^L_{\widetilde{A}_1}, \gamma^U_{\widetilde{A}_1}] \in u_{\widetilde{A}_1}, [\gamma^L_{\widetilde{A}_2}, \gamma^U_{\widetilde{A}_2}] \in u_{\widetilde{A}_2}} \{[\gamma^L_{\widetilde{A}_1} \times \gamma^L_{\widetilde{A}_2}, \\ & \quad \gamma^U_{\widetilde{A}_1} \times \gamma^U_{\widetilde{A}_2}]\}, \end{aligned}$$

$$\bigcup_{\left[\delta^L_{\tilde{A}_1},\delta^U_{\tilde{A}_1}\right]\in w_{\tilde{A}_1},\left[\delta^L_{\tilde{A}_2},\delta^U_{\tilde{A}_1}\right]\in w_{\tilde{A}_2}} \{[\delta^L_{\tilde{A}_1}+\delta^L_{\tilde{A}_2}-\delta^L_{\tilde{A}_1}\times\delta^L_{\tilde{A}_2},$$
$$\delta^U_{\tilde{A}_1}+\delta^U_{\tilde{A}_2}-\delta^U_{\tilde{A}_1}\times\delta^U_{\tilde{A}_2}]\},$$
$$\bigcup_{\left[\eta^L_{\tilde{A}_1},\eta^U_{\tilde{A}_1}\right]\in v_{\tilde{A}_1},\left[\eta^L_{\tilde{A}_2},\eta^U_{\tilde{A}_2}\right]\in v_{\tilde{A}_2}} \{[\eta^L_{\tilde{A}_1}+\eta^L_{\tilde{A}_2}-\eta^L_{\tilde{A}_1}\times\eta^L_{\tilde{A}_2},$$
$$\eta^U_{\tilde{A}_1}+\eta^U_{\tilde{A}_2}-\eta^U_{\tilde{A}_1}\times\eta^U_{\tilde{A}_2}]\});$$
$$(11.14)$$

$$\lambda h_{\tilde{A}} = \Big(\bigcup_{\left[\gamma^L_{\tilde{A}},\gamma^U_{\tilde{A}}\right]\in u_{\tilde{A}}} \{1-(1-\gamma^L_{\tilde{A}})^\lambda,1-(1-\gamma^U_{\tilde{A}})^\lambda]\},$$
$$\bigcup_{\left[\delta^L_{\tilde{A}},\delta^U_{\tilde{A}}\right]\in w_{\tilde{A}}} \{[(\delta^L_{\tilde{A}})^\lambda,(\delta^U_{\tilde{A}})^\lambda]\},$$
$$\bigcup_{\left[\eta^L_{\tilde{A}},\eta^U_{\tilde{A}}\right]\in v_{\tilde{A}}} \{[(\eta^L_{\tilde{A}})^\lambda,(\eta^U_{\tilde{A}})^\lambda]\}); \tag{11.15}$$

$$h_{\tilde{A}}^{\lambda} = \Big(\bigcup_{\left[\gamma^L_{\tilde{A}},\gamma^U_{\tilde{A}}\right]\in u_{\tilde{A}}} \{[(\gamma^L_{\tilde{A}})^\lambda,(\gamma^U_{\tilde{A}})^\lambda]\},$$
$$\bigcup_{\left[\delta^L_{\tilde{A}},\delta^U_{\tilde{A}}\right]\in w_{\tilde{A}}} \{[1-(1-\delta^L_{\tilde{A}})^\lambda,1-(1-\delta^U_{\tilde{A}})^\lambda]\},$$
$$\bigcup_{\left[\eta^L_{\tilde{A}},\eta^U_{\tilde{A}}\right]\in v_{\tilde{A}}} \{[1-(1-\eta^L_{\tilde{A}})^\lambda,1-(1-\eta^U_{\tilde{A}})^\lambda]\}). \tag{11.16}$$

Remark 11.3 There are also cases reported in the literature, in which hesitant picture fuzzy linguistic set [8] and its revision form [9] have been taken into consideration.

References

1. R. Ambrin, M. Ibrar, M. De La Sen, I. Rabbi, A. Khan, Extended TOPSIS method for supplier selection under picture hesitant fuzzy environment using linguistic variables. J. Math. Article ID 6652586. **2021** (2021)
2. B. Cuong, Picture fuzzy sets-first results, Part 1, in *Proceedings of the Third World Congress on Information and Communication WICT'2013*, Hanoi, Vietnam, (15–18 December 2013), pp. 1–6
3. N. Jan, Z. Ali, T. Mahmood, K. Ullah, Some generalized distance and similarity measures for picture hesitant fuzzy sets and their applications in building material recognition and multi-attribute decision making. Punjab Univ. J. Math. **51**, 51–70 (2019)
4. T. Mahmood, Z. Ali, The fuzzy cross-entropy for picture hesitant fuzzy sets and their application in multi criteria decision making. Punjab Univ. J. Math. **52**, 55–82 (2020)
5. W. Ullah, M. Ibrar, A. Khan, M. Khan, Multiple attribute decision making problem using GRA method with incomplete weight information based on picture hesitant fuzzy setting. Int. J. Intell. Syst. **36**, 866–889 (2021)
6. R. Wang, Y. Li, Picture hesitant fuzzy set and its application to multiple criteria decision-making. Symmetry **10**, 295 (2018)
7. G.W. Wei, Picture fuzzy aggregation operators and their application to multiple attribute decision making. J. Intell. Fuzzy Syst. **33**, 713–724 (2017)
8. L. Yang, X. Wu, J. Qian, A novel multicriteria group decision making approach with hesitant picture fuzzy linguistic information. Math. Probl. Eng. Article ID 6394028. **2020** (2020)
9. X. Zhang, J. Wang, J. Wang, J. Hu, A revised picture fuzzy linguistic aggregation operator and its application to group decision-making. Cognit. Computat. **12**, 1070–1082 (2020)

Chapter 12
Spherical Hesitant Fuzzy Set

Abstract In this chapter, we present the concept of spherical hesitant fuzzy set as a generalization of both picture hesitant fuzzy set and Pythagorean hesitant fuzzy set for dealing with fuzziness and uncertainty information. In the sequel, we are going to re-state the extension form of spherical hesitant fuzzy set by generalizing that to the concept of T-spherical hesitant fuzzy set.

12.1 Spherical Hesitant Fuzzy Set

By keeping the importance of logarithmic function of spherical hesitant fuzzy set (SHFS) and its aggregation operators in mind, Khan et al. [3] developed an algorithm to tackle the multiple criteria decision making under spherical hesitant fuzzy environment. Naeem et al. [4] proposed a number of aggregation operations for SHFSs by the use of sine trigonometric function, namely the spherical hesitant fuzzy weighted average and the spherical hesitant fuzzy weighted geometric aggregation operators.

By keeping the reference set as X in mind, a spherical fuzzy set A on X is defined [2] in terms of three functions $u_A(x)$, $w_A(x)$, and $v_A(x)$ as follows:

$$A = \{\langle x, u_A(x), w_A(x), v_A(x)\rangle \mid x \in X\}, \tag{12.1}$$

where $u_A(x)$, $w_A(x)$, and $v_A(x)$ are the sets of some different values in $[0, 1]$ and represent the positive, neutral, and negative membership degrees of the element $x \in X$ to A, respectively. The positive, neutral, and negative membership degrees satisfy

$$0 \leq (u_A(x))^2 + (w_A(x))^2 + (v_A(x))^2 \leq 1 \tag{12.2}$$

B. Farhadinia, *Hesitant Fuzzy Set*, Computational Intelligence Methods and Applications, https://doi.org/10.1007/978-981-16-7301-6_12

and make a presentation of indeterminacy degree as $\pi_A(x) = \sqrt[2]{1-(u_A(x))^2-(w_A(x))^2-(v_A(x))^2}$ for any $x \in X$. It is easily observed that the concept of spherical fuzzy set is an extension of picture fuzzy set.

Given a fixed $x \in X$, some operational laws on spherical fuzzy sets are defined as follows (see [3]):

$$\begin{aligned}
A^c &= \{\langle x, v_A(x), w_A(x), u_A(x)\rangle \mid x \in X\};\\
A_1 \cup A_2 &= \{\langle x, \max\{u_{A_1}(x), u_{A_2}(x)\}, \min\{w_{A_1}(x), w_{A_2}(x)\},\\
&\quad \min\{v_{A_1}(x), v_{A_2}(x)\}\rangle \mid x \in X\};\\
A_1 \cap A_2 &= \{\langle x, \min\{u_{A_1}(x), u_{A_2}(x)\}, \max\{w_{A_1}(x), w_{A_2}(x)\},\\
&\quad \max\{v_{A_1}(x), v_{A_2}(x)\}\rangle \mid x \in X\};\\
A_1 \oplus A_2 &= \{\langle x, u_{A_1}(x) + u_{A_2}(x) - u_{A_1}(x) \times u_{A_2}(x), w_{A_1}(x) \times w_{A_2}(x),\\
&\quad v_{A_1}(x) \times v_{A_2}(x)\}\rangle \mid x \in X\};\\
A_1 \otimes A_2 &= \{\langle x, u_{A_1}(x) \times u_{A_2}(x), w_{A_1}(x) + w_{A_2}(x) - w_{A_1}(x) \times w_{A_2}(x),\\
&\quad v_{A_1}(x) + v_{A_2}(x) - v_{A_1}(x) \times v_{A_2}(x)\}\rangle \mid x \in X\};\\
\lambda A &= \{\langle x, 1-(1-u_A(x))^\lambda, (w_A(x))^\lambda, (v_A(x))^\lambda\rangle \mid x \in X\};\\
A^\lambda &= \{\langle x, (u_A(x))^\lambda, 1-(1-w_A(x))^\lambda, 1-(1-v_A(x))^\lambda\rangle \mid x \in X\},
\end{aligned}$$

where $\lambda > 0$, and

$$\begin{aligned}
&0 \le (u_A(x))^2 + (w_A(x))^2 + (v_A(x))^2 \le 1,\\
&0 \le (u_{A_1}(x))^2 + (w_{A_1}(x))^2 + (v_{A_1}(x))^2 \le 1,\\
&0 \le (u_{A_2}(x))^2 + (w_{A_2}(x))^2 + (v_{A_2}(x))^2 \le 1.
\end{aligned}$$

Example 12.1 Let $X = \{x_1, x_2\}$ be the reference set.
The two sets $h_A(x_1) = \{(0.2, 0.2, 0.3)\}$ and $h_A(x_2) = \{(0.4, 0.1, 0.3)\}$ are spherical fuzzy sets of x_i $(i = 1, 2)$ belonging to a set A, and this is clear from the fact that $0.2^2 + 0.2^2 + 0.3^2 \le 1$ and $0.4^2 + 0.1^2 + 0.3^2 \le 1$. Thus A can be considered as a spherical fuzzy set, i.e.,

$$A = \{\langle x_1, (0.2, 0.2, 0.3)\rangle, \ \langle x_2, (0.4, 0.1, 0.3)\rangle\}.$$

Definition 12.1 ([4]) Let X be a reference set. A spherical hesitant fuzzy set (SHFS) A on X is defined in terms of three functions $u_A(x)$, $w_A(x)$ and $v_A(x)$ as follows:

$$A = \{\langle x, u_A(x), w_A(x), w_A(x)\rangle \mid x \in X\}, \tag{12.3}$$

where $u_A(x)$, $w_A(x)$, and $v_A(x)$ are the sets of some different values in [0, 1] and represent the possible positive, neutral, and negative membership degrees of the element $x \in X$ to A, respectively.

Here, for all $x \in X$, if we take $u_A(x) = \bigcup_{\gamma_A \in u_A(x)}\{\gamma_A\}$, $w_A(x) = \bigcup_{\delta_A \in w_A(x)}\{\delta_A\}$, $v_A(x) = \bigcup_{\eta_A \in v_A(x)}\{\eta_A\}$, $\gamma_A^+ \in u_A^+ = \bigcup_{x\in X} \max_{\gamma_A \in u_A(x)}\{\gamma_A\}$, $\delta_A^+ \in w_A^+ = \bigcup_{x\in X} \max_{\delta_A \in w_A(x)}\{\delta_A\}$ and $\eta_A^+ \in v_A^+ = \bigcup_{x\in X} \max_{\eta_A \in v_A(x)}\{\eta_A\}$, then we conclude that

$$0 \le \gamma_A,\ \delta_A,\ \eta_A \le 1, \quad 0 \le (\gamma_A^+)^2 + (\delta_A^+)^2 + (\eta_A^+)^2 \le 1.$$

For the sake of simplicity, the triple $h_A(x) = (u_A(x), w_A(x), v_A(x))$ is called the spherical hesitant fuzzy element (SHFE).

Remark 12.1 Throughout this book, the set of all SHFSs on the reference set X is denoted by $\mathbb{SHFS}(X)$.

Example 12.2 Let $X = \{x_1, x_2\}$ be the reference set, $h_A(x_1) = (u_A(x_1), w_A(x_1), v_A(x_1)) = (\{0.2, 0.3\}, \{0.1, 0.4\}, \{0.3\})$ and $h_A(x_2) = (u_A(x_2), w_A(x_2), v_A(x_2)) = (\{0.3, 0.4\}, \{0.2\}, \{0.1, 0.2\})$. Then, $h_A(x_i)$ for $i = 1, 2$ are the SHFEs of x_i $(i = 1, 2)$ in the set A, because the relations

$$\gamma_A^+(x_1) = 0.3,\ \delta_A^+(x_1) = 0.4,\ \eta_A^+(x_1) = 0.3;$$
$$\gamma_A^+(x_2) = 0.4,\ \delta_A^+(x_2) = 0.2,\ \eta_A^+(x_2) = 0.2$$

result in

$$(\gamma_A^+)^2(x_1) + (\delta_A^+)^2(x_1) + (\eta_A^+)^2(x_1) = 0.3^2 + 0.4^2 + 0.3^2 \le 1;$$
$$(\gamma_A^+)^2(x_2) + (\delta_A^+)^2(x_2) + (\eta_A^+)^2(x_2) = 0.4^2 + 0.2^2 + 0.2^2 \le 1.$$

Thus, A is a SHFS, and it is denoted by

$$A = \{\langle x_1, (\{0.2, 0.3\}, \{0.1, 0.4\}, \{0.3\})\rangle,\ \langle x_2, (\{0.3, 0.4\}, \{0.2\}, \{0.1, 0.2\})\rangle\}.$$

For the SHFEs $h_A = (u_A, w_A, v_A)$, $h_{A_1} = (u_{A_1}, w_{A_1}, v_{A_1})$, and $h_{A_2} = (u_{A_2}, w_{A_2}, v_{A_2})$ the following operations are defined:

$$(h_A)^c = (v_A, w_A, u_A); \tag{12.4}$$
$$h_{A_1} \cup h_{A_2} = (u_{A_1} \cup u_{A_2}, w_{A_1} \cap w_{A_2}, v_{A_1} \cap v_{A_2}); \tag{12.5}$$
$$h_{A_1} \cap h_{A_2} = (u_{A_1} \cap u_{A_2}, w_{A_1} \cup w_{A_2}, v_{A_1} \cup v_{A_2}); \tag{12.6}$$
$$h_{A_1} \oplus h_{A_2} = (u_{A_1} \oplus u_{A_2}, w_{A_1} \otimes w_{A_2}, v_{A_1} \otimes v_{A_2}); \tag{12.7}$$
$$h_{A_1} \otimes h_{A_2} = (u_{A_1} \otimes u_{A_2}, w_{A_1} \oplus w_{A_2}, v_{A_1} \oplus v_{A_2}), \tag{12.8}$$

where

$$0 \leq (\gamma_A^+)^2 + (\delta_A^+)^2 + (\eta_A^+)^2 \leq 1,$$
$$0 \leq (\gamma_{A_1}^+)^2 + (\delta_{A_1}^+)^2 + (\eta_{A_1}^+)^2 \leq 1,$$
$$0 \leq (\gamma_{A_2}^+)^2 + (\delta_{A_2}^+)^2 + (\eta_{A_2}^+)^2 \leq 1.$$

On the basis of the above operations on SHFEs, some relationships can be further established for such operations on SHFSs as follows:

$$A_1 \cup A_2 = \bigcup_{h_{A_1} \in A_1, h_{A_2} \in A_2} h_{A_1} \cup h_{A_2}; \tag{12.9}$$

$$A_1 \cap A_2 = \bigcup_{h_{A_1} \in A_1, h_{A_2} \in A_2} h_{A_1} \cap h_{A_2}; \tag{12.10}$$

$$A_1 \oplus A_2 = \bigcup_{h_{A_1} \in A_1, h_{A_2} \in A_2} h_{A_1} \oplus h_{A_2}; \tag{12.11}$$

$$A_1 \otimes A_2 = \bigcup_{h_{A_1} \in A_1, h_{A_2} \in A_2} h_{A_1} \otimes h_{A_2}. \tag{12.12}$$

12.2 T-Spherical Hesitant Fuzzy Set

Naeem et al. [4] presented the notion of T-spherical hesitant fuzzy set (TSHFS), and they enhanced the topic with introducing basic operations for TSHFSs using sine trigonometric function. Furthermore, they proposed some aggregation operators for TSHFSs which are implemented in the multiple criteria decision making in the spherical hesitant fuzzy information.

Definition 12.2 ([4]) Let X be a reference set. A T-spherical hesitant fuzzy set (T-SHFS) $\widetilde{A}$ on X is defined in terms of three functions $u_{\widetilde{A}}(x)$, $w_{\widetilde{A}}(x)$, and $v_{\widetilde{A}}(x)$ as follows:

$$\widetilde{A} = \{\langle x, u_{\widetilde{A}}(x), w_{\widetilde{A}}(x), w_{\widetilde{A}}(x)\rangle \mid x \in X\}, \tag{12.13}$$

where $u_{\widetilde{A}}(x)$, $w_{\widetilde{A}}(x)$, and $v_{\widetilde{A}}(x)$ are the sets of some different values in [0, 1] and represent the possible positive, neutral, and negative membership degrees of the element $x \in X$ to $\widetilde{A}$, respectively.

Here, for all $x \in X$, if we consider $u_{\widetilde{A}}(x) = \bigcup_{\gamma_{\widetilde{A}} \in u_{\widetilde{A}}(x)}\{\gamma_{\widetilde{A}}\}$, $w_{\widetilde{A}}(x) = \bigcup_{\delta_{\widetilde{A}} \in w_{\widetilde{A}}(x)}\{\delta_{\widetilde{A}}\}$, $v_{\widetilde{A}}(x) = \bigcup_{\eta_{\widetilde{A}} \in v_{\widetilde{A}}(x)}\{\eta_{\widetilde{A}}\}$, $\gamma^+_{\widetilde{A}} \in u^+_{\widetilde{A}} = \bigcup_{x \in X} \max_{\gamma_{\widetilde{A}} \in u_{\widetilde{A}}(x)}\{\gamma_{\widetilde{A}}\}$, $\delta^+_{\widetilde{A}} \in w^+_{\widetilde{A}} = \bigcup_{x \in X} \max_{\delta_{\widetilde{A}} \in w_{\widetilde{A}}(x)}\{\delta_{\widetilde{A}}\}$ and $\eta^+_{\widetilde{A}} \in v^+_{\widetilde{A}} = \bigcup_{x \in X} \max_{\eta_{\widetilde{A}} \in v_{\widetilde{A}}(x)}\{\eta_{\widetilde{A}}\}$, then we find that

$$0 \leq \gamma_{\widetilde{A}},\ \delta_{\widetilde{A}},\ \eta_{\widetilde{A}} \leq 1, \quad 0 \leq (\gamma^+_{\widetilde{A}})^q + (\delta^+_{\widetilde{A}})^q + (\eta^+_{\widetilde{A}})^q \leq 1, \quad q \geq 1.$$

For the sake of simplicity, the triple $h_{\widetilde{A}}(x) = (u_{\widetilde{A}}(x), w_{\widetilde{A}}(x), v_{\widetilde{A}}(x))$ is called the T-spherical hesitant fuzzy element (T-SHFE).

Remark 12.2 Throughout this book, the set of all T-SHFSs on the reference set X is denoted by $\mathbb{T} - \mathbb{SHFS}(X)$.

Example 12.3 Let $X = \{x_1, x_2\}$ be the reference set, $h_{\widetilde{A}}(x_1) = (u_{\widetilde{A}}(x_1), w_{\widetilde{A}}(x_1), v_{\widetilde{A}}(x_1)) = (\{0.2, 0.3\}, \{0.1, 0.4\}, \{0.3\})$ and $h_{\widetilde{A}}(x_2) = (u_{\widetilde{A}}(x_2), w_{\widetilde{A}}(x_2), v_{\widetilde{A}}(x_2)) = (\{0.3, 0.4\}, \{0.2\}, \{0.1, 0.2\})$. Then, $h_{\widetilde{A}}(x_i)$ for $i = 1, 2$ are the T-SHFEs of x_i $(i = 1, 2)$ in the set $\widetilde{A}$, because the relations

$$\gamma^+_A(x_1) = 0.3,\ \delta^+_A(x_1) = 0.4,\ \eta^+_A(x_1) = 0.3;$$
$$\gamma^+_A(x_2) = 0.4,\ \delta^+_A(x_2) = 0.2,\ \eta^+_A(x_2) = 0.2$$

result in

$$(\gamma^+_A)^q(x_1) + (\delta^+_A)^q(x_1) + (\eta^+_A)^q(x_1) = 0.3^q + 0.4^q + 0.3^q \leq 1;$$
$$(\gamma^+_A)^q(x_2) + (\delta^+_A)^q(x_2) + (\eta^+_A)^q(x_2) = 0.4^q + 0.2^q + 0.2^q \leq 1, \quad q \geq 1.$$

Thus, $\widetilde{A}$ is a T-SHFS, and it is denoted by

$$\widetilde{A} = \{\langle x_1, (\{0.2, 0.3\}, \{0.1, 0.4\}, \{0.3\})\rangle,\ \langle x_2, (\{0.3, 0.4\}, \{0.2\}, \{0.1, 0.2\})\rangle\}.$$

For the T-SHFEs $h_{\widetilde{A}} = (u_{\widetilde{A}}, w_{\widetilde{A}}, v_{\widetilde{A}})$, $h_{\widetilde{A}_1} = (u_{\widetilde{A}_1}, w_{\widetilde{A}_1}, v_{\widetilde{A}_1})$ and $h_{\widetilde{A}_2} = (u_{\widetilde{A}_2}, w_{\widetilde{A}_2}, v_{\widetilde{A}_2})$ the following operations are defined:

$$(h_{\widetilde{A}})^c = (v_{\widetilde{A}}, w_{\widetilde{A}}, u_{\widetilde{A}}); \tag{12.14}$$
$$h_{\widetilde{A}_1} \cup h_{\widetilde{A}_2} = (u_{\widetilde{A}_1} \cup u_{\widetilde{A}_2}, w_{\widetilde{A}_1} \cap w_{\widetilde{A}_2}, v_{\widetilde{A}_1} \cap v_{\widetilde{A}_2}); \tag{12.15}$$
$$h_{\widetilde{A}_1} \cap h_{\widetilde{A}_2} = (u_{\widetilde{A}_1} \cap u_{\widetilde{A}_2}, w_{\widetilde{A}_1} \cup w_{\widetilde{A}_2}, v_{\widetilde{A}_1} \cup v_{\widetilde{A}_2}); \tag{12.16}$$
$$h_{\widetilde{A}_1} \oplus h_{\widetilde{A}_2} = (u_{\widetilde{A}_1} \oplus u_{\widetilde{A}_2}, w_{\widetilde{A}_1} \otimes w_{\widetilde{A}_2}, v_{\widetilde{A}_1} \otimes v_{\widetilde{A}_2}); \tag{12.17}$$
$$h_{\widetilde{A}_1} \otimes h_{\widetilde{A}_2} = (u_{\widetilde{A}_1} \otimes u_{\widetilde{A}_2}, w_{\widetilde{A}_1} \oplus w_{\widetilde{A}_2}, v_{\widetilde{A}_1} \oplus v_{\widetilde{A}_2}), \tag{12.18}$$

where

$$0 \leq (\gamma_{\widetilde{A}}^{+})^q + (\delta_{\widetilde{A}}^{+})^q + (\eta_{\widetilde{A}}^{+})^q \leq 1,$$
$$0 \leq (\gamma_{\widetilde{A}_1}^{+})^q + (\delta_{\widetilde{A}_1}^{+})^q + (\eta_{\widetilde{A}_1}^{+})^q \leq 1,$$
$$0 \leq (\gamma_{\widetilde{A}_2}^{+})^q + (\delta_{\widetilde{A}_2}^{+})^q + (\eta_{\widetilde{A}_2}^{+})^q \leq 1, \quad q \geq 1.$$

On the basis of the above operations on T-SHFEs, some relationships can be further established for such operations on T-SHFSs as follows:

$$\widetilde{A}_1 \cup \widetilde{A}_2 = \bigcup_{h_{\widetilde{A}_1} \in \widetilde{A}_1, h_{\widetilde{A}_2} \in \widetilde{A}_2} h_{\widetilde{A}_1} \cup h_{\widetilde{A}_2}; \tag{12.19}$$

$$\widetilde{A}_1 \cap \widetilde{A}_2 = \bigcup_{h_{\widetilde{A}_1} \in \widetilde{A}_1, h_{\widetilde{A}_2} \in \widetilde{A}_2} h_{\widetilde{A}_1} \cap h_{\widetilde{A}_2}; \tag{12.20}$$

$$\widetilde{A}_1 \oplus \widetilde{A}_2 = \bigcup_{h_{\widetilde{A}_1} \in \widetilde{A}_1, h_{\widetilde{A}_2} \in \widetilde{A}_2} h_{\widetilde{A}_1} \oplus h_{\widetilde{A}_2}; \tag{12.21}$$

$$\widetilde{A}_1 \otimes \widetilde{A}_2 = \bigcup_{h_{\widetilde{A}_1} \in \widetilde{A}_1, h_{\widetilde{A}_2} \in \widetilde{A}_2} h_{\widetilde{A}_1} \otimes h_{\widetilde{A}_2}. \tag{12.22}$$

Remark 12.3 Also, we are able to get other extensions for spherical (or T-spherical) hesitant fuzzy set by combining the concepts declared in this book. For instance, the concept of complex spherical fuzzy set [1] is created by mixing the concepts of complex fuzzy set and spherical fuzzy set.

References

1. Z. Ali, T. Mahmood, M. Yang, Complex T-spherical fuzzy aggregation operators with application to multi-attribute decision making. Symmetry **12**, 1311 (2020)
2. S. Ashraf, S. Abdullah, T. Mahmood, GRA method based on spherical linguistic fuzzy Choquet integral environment and its application in multi-attribute decision-making problems. Math. Sci. **12**, 263–275 (2018)
3. A. Khan, S. Abosuliman, S. Abdullah, M. Ayaz, A decision support model for hotel recommendation based on the online consumer reviews using logarithmic spherical hesitant fuzzy information. Entropy **23**, 432 (2021)
4. M. Naeem, A. Khan, S. Abdullah, S. Ashraf, A. Khammash, Solid waste collection system selection based on sine trigonometric spherical hesitant fuzzy aggregation information. Intell. Autom. Soft Comput. **28**, 459–476 (2021)

Index

B. Farhadinia, *Hesitant Fuzzy Set*, Computational Intelligence Methods and Applications, https://doi.org/10.1007/978-981-16-7301-6

T

Printed in the United States
by Baker & Taylor Publisher Services